Industrial Hygiene in the Pharmaceutical and Consumer Healthcare Industries

This volume is an update on the use of containment in the pharmaceutical industry and consumer healthcare. It serves to highlight how industrial hygiene acts as a driving force within these industries to reduce the risk of exposure to chemical and physical agents, particularly to powders and dusts, while taking all factors into account. The author emphasizes how this book is not designed to replace other texts on containment; rather, it will serve to show a practical approach of utilizing the technologies within the high-demand industries of pharmaceuticals and consumer healthcare.

Features

- Timely coverage of changes in process control technology for the pharmaceutical industry, a dynamic area in terms of products and manufacturing processes
- Provides an update on the unique requirements of these industries and how they differ from others, for example the microelectronics or specialized chemicals industries
- Draws on the author's vast experience in the field of industrial hygiene and hazardous materials
- Presents a collection of unique situations in which industrial hygiene was implemented to resolve a variety of scenarios and did not interfere with quality issues
- Addresses current topics relating to industry evolution such as migration of therapies to higher potency, RiskMAP, new modalities in medicines and treatments, and large-molecule therapeutics and conjugates

Drugs and the Pharmaceutical Sciences

A Series of Textbooks and Monographs

Series Editor

Anthony J. Hickey

RTI International, Research Triangle Park, USA

The Drugs and Pharmaceutical Sciences series is designed to enable the pharmaceutical scientist to stay abreast of the changing trends, advances and innovations associated with therapeutic drugs and that area of expertise and interest that has come to be known as the pharmaceutical sciences. The body of knowledge that those working in the pharmaceutical environment have to work with, and master, has been, and continues, to expand at a rapid pace as new scientific approaches, technologies, instrumentations, clinical advances, economic factors and social needs arise and influence the discovery, development, manufacture, commercialization and clinical use of new agents and devices.

Percutaneous Absorption: Drugs, Cosmetics, Mechanisms, Methods, Nina Dragićević and Howard Maibach

Handbook of Pharmaceutical Granulation Technology, Dilip M. Parikh

Biotechnology: the Science, the Products, the Government, the Business

Emerging Drug Delivery and Biomedical Engineering Technologies: Transforming Therapy, Dimitrios Lamprou

RNA-seq in Drug Discovery and Development, Feng Cheng and Robert Morris

Patient Safety in Developing Countries: Education, Research, Case Studies, Yaser Al-Worafi

Industrial Hygiene in the Pharmaceutical and Consumer Healthcare Industries, Casey Cosner

Cancer Targeting Therapies: Conventional and Advanced Perspectives, Muhammad Yasir Ali, Shazia Bukhari

For more information about this series, please visit: www.crcpress.com/Drugs-and-the-Pharmaceutical-Sciences/book-series/IHCDRUPHASCI

Industrial Hygiene in the Pharmaceutical and Consumer Healthcare Industries

Authored By

Casey Cosner

EHS Adviser – Industrial Hygiene, GlaxoSmithKline
LLC 320
South Broadway St. Louis, MO 63102

CRC Press
Taylor & Francis Group
Boca Raton London New York

CRC Press is an imprint of the
Taylor & Francis Group, an **informa** business

First edition published 2024
by CRC Press
6000 Broken Sound Parkway NW, Suite 300, Boca Raton, FL 33487-2742

and by CRC Press
4 Park Square, Milton Park, Abingdon, Oxon, OX14 4RN

© 2024 Casey Cosner

CRC Press is an imprint of Taylor & Francis Group, LLC

Library of Congress Cataloging-in-Publication Data
Names: Cosner, Casey, author.
Title: Industrial hygiene in the pharmaceutical and consumer healthcare industries / Casey Cosner.
Description: Boca Raton : CRC Press, 2024. | Series: Drugs and the pharmaceutical sciences | Includes bibliographical references and index. | Summary: "This volume is an update on the use of containment in the pharmaceutical industry and consumer healthcare. It serves to highlight how industrial hygiene acts as a driving force within these industries to reduce the risk of exposure to chemical and physical agents, particularly to powders and dusts, while taking all factors into account. The author emphasizes how this book is not designed to replace other texts on containment; rather, it will serve to show a practical approach of utilizing the technologies within the high-demand industries of pharmaceuticals and consumer healthcare"– Provided by publisher.
Identifiers: LCCN 2023020496 (print) | LCCN 2023020497 (ebook) | ISBN 9781032226309 (hardback) | ISBN 9781032226354 (paperback) | ISBN 9781003273455 (ebook)
Subjects: LCSH: Pharmaceutical industry–United States–Safety measures. | Pharmaceutical industry–Risk management–United States. | Drug factories–United States–Safety measures. | Drug factories–Risk assessment–United States.
Classification: LCC HD9666.5. C666 2024 (print) | LCC HD9666.5 (ebook) | DDC 338.4/761510973–dc23/eng/20230802
LC record available at https://lccn.loc.gov/2023020496
LC ebook record available at https://lccn.loc.gov/2023020497

ISBN: 978-1-032-2-2630-9 (HB)
ISBN: 978-1-032-2-2635-4 (PB)
ISBN: 978-1-003-2-7345-5 (EB)

DOI: 10.1201/9781003273455

Typeset in Times
by codeMantra

This book is dedicated to Amberlyn, Carolyn, Ethan, and Ryan. You are all a daily inspiration, and this project would not have been completed without your encouragement.

Contents

Preface

Industrial hygiene (often known as occupational hygiene outside the United States) is the art and science of anticipation, recognition, evaluation, and control of workplace hazards to employees. The pharmaceutical industry in particular is rife with hazards, their processes and products chief among them. Beginning in the 1980s, recognizing that its products were becoming more potent and dangerous, the industry began to police itself and utilize industrial hygiene as a means to protect its single greatest asset: the employees. The industry took a more stringent approach toward worker and health and safety by crafting conservative occupational exposure limits (OELs), requiring better sampling methods and more stringent data analyses with less uncertainty. But the industry then expanded the utility of industrial hygiene even further, asking the practitioners of the craft to aid in developing better tools to protect the workers. And in the modern era, those same organizations are demanding the same requirements but in an environmentally sustainable manner. The industry as a whole has been practicing industrial hygiene through the lens of risk management.

Largely based on the ISO 31000 Risk Management Standard, the field of industrial hygiene in the pharmaceutical industry is highly regimented and aligns with the concepts of risk assessment, risk characterization, and risk treatment. The undeniable value that is wrought from the power of industrial hygiene is brought forth in this manner and forces the practicing industrial hygienist to be well versed in numerous fields. This book is an overview of how the modern pharmaceutical industry utilizes industrial hygiene to keep its employees safe while handling what may be the broadest collection of chemical hazards on the planet. It is not intended to be a pedagogical tool on how to do industrial hygiene; rather, it is intended to showcase the approach taken by the industry and the significant number of challenges that the practicing hygienist must face while performing a project.

Importantly, this book focuses exclusively on evaluation and control of airborne exposures. While the field of industrial hygiene encompasses many more hazardous stressors, including noise, thermal, ergonomics, radiation, and more, for the sake of brevity, this book relegates the scope to airborne contaminants. However, the same approach is utilized within the industry for these additional stressors as well. Finally, the approach that is articulated in this book and championed by the pharmaceutical industry is not strictly applicable to the industry. It can (and should!) be applied to essentially any industry, but in particular those that are particularly chemical-heavy, such as the chemical manufacturing industry, perfumes and fragrances, flavorings, spices, and dyes.

Completing this book was a labor of love, and I sincerely hope that you gain a better understanding of how the industry utilizes industrial hygiene and the value that industrial hygienists can bring to the table. Despite my greatest efforts, it is almost certain that I left details out. Perhaps with luck future editions can expound on some of these topics further.

Author

Dr Casey C. Cosner, PhD, CIH, CHMM, is the Industrial Hygiene and Occupational Health Manager for MilliporeSigma where he oversees all aspects of the industrial hygiene program.He has worked for and performed industrial hygiene projects within the pharmaceutical, consumer healthcare, contract manufacturing, professional, and biomedical research settings. He is a Certified Industrial Hygienist (CIH) and a Certified Hazardous Materials Manager (CHMM). He is an active member of the American Industrial Hygiene Association (AIHA). Casey earned his Bachelor of Science in Chemistry from the University of South Florida and obtained his PhD in Chemistry from the University of Notre Dame, specializing in synthetic organic chemistry.

1 Background and Introduction

1.1 THE IMPACT OF PHARMACEUTICAL COMPANIES

The pharmaceutical industry is one of the most complex and impactful industries ever created. A few other industries have combined scientific curiosity and engineering ingenuity to solve problems plaguing mankind with as much impact as the pharmaceutical industry has. The creation and distribution of numerous life-saving medicines have significantly contributed to increasing the average life span of people worldwide. In the United States alone, between 1860 and 2020, the average life expectancy rose from 39.4 to 79.9 years (Figure 1.1). While other factors most certainly contributed to this drastic increase (such as readily available clean water, increased food supply, enhanced public infrastructure, healthier lifestyles and advances in medical science), the impact of the pharmaceutical industry cannot be overstated.

The demand for pharmaceuticals continues to increase as well. Figure 1.2 shows the global annual revenue of the pharmaceutical industry from 2001 to 2020. In that 20-year span, the revenue for the industry as a whole increased by 324% to a total of $1.2 trillion. While some of that increase can be attributed to the rising cost of prescription drugs, the overall demand is a major driving force. Within that same time frame, the number of prescriptions issued to patients increased every year. Clearly, there is significant demand and need for prescription drugs, and it does not seem to be in any danger of abating. In order to meet this increasing demand, pharmaceutical sites are manufacturing more materials and at a greater pace than ever before.

Likewise, the consumer healthcare sector is also experiencing similar trends. A branch of the pharmaceutical industry that is often overlooked, consumer healthcare products typically include common over-the-counter (OTC) medications such as cold and flu remedies, pain relief, digestive aids, eye drops, lotions and ointments, supplements, and oral care, just to name a few. Like with prescription drugs, these products are designed to illicit some physiological effect at the administered

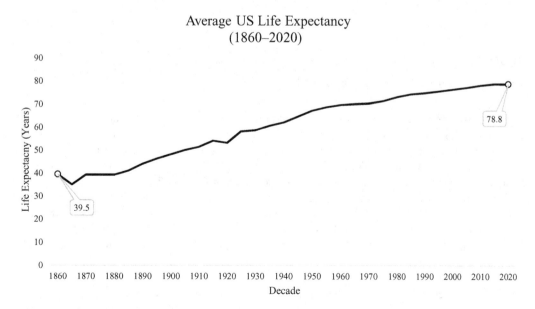

FIGURE 1.1   The increase in United States life expectancy from 1860 to 2020.

Source: Statista.

DOI: 10.1201/9781003273455-1

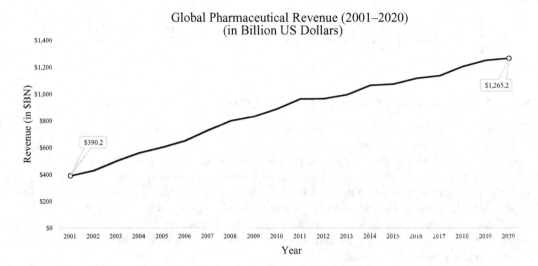

FIGURE 1.2   The global pharmaceutical industry revenue from 2001 to 2020.

Source: Statista.

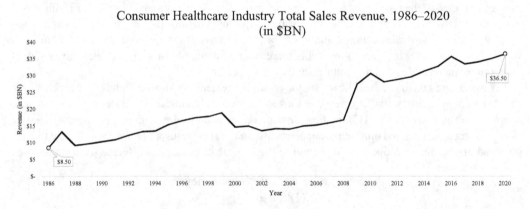

FIGURE 1.3   The global revenue for the consumer healthcare industry from 1986 to 2020.

Source: Consumer Healthcare Products Association.

dose but have been deemed generally safe enough for the general population that a prescription is not required. Figure 1.3 shows the increase in consumer healthcare revenue between 1986 and 2020. During this time period, the overall trend for the industry was a 429% increase in revenue from sales within the United States alone. (Between 2000 and 2009, the official definition of "consumer healthcare product" changed, accounting for the noted dip in Figure 1.3.)

While the consumer healthcare raw revenue number is not nearly as impressive as that for the prescription drugs ($1.2 trillion vs. $36.5 billion), the reader should note that there is a larger percentage increase for consumer healthcare products. Over the last decade, more and more consumers are turning to OTC products to manage their ailments. The pandemic of 2019 and 2020 only served to exacerbate these trends even more, causing an array of supply chain issues. The overall trend in shifting toward OTC products by consumers is not expected to plateau either, as consumers continue to wrestle with rising drug costs and the complexities of health insurance and with scheduling doctor visits.

To meet these increasing demands for both prescription and OTC products, parent companies must rely on their employees to keep all the manufacturing sites operational. As mentioned above, the materials being manufactured and handled are often designed to be biologically active at small doses, such is the case with active ingredients in the products. Other materials, most notably excipients, are not intended to have biological effects but may unintentionally exhibit some effects at higher doses. Both classes of substances are handled in large quantities in both manufacturing sites (primary pharmaceutical sites) and packaging sites (secondary pharmaceutical sites). The routine handling of such materials may result in the unintentional exposure of employees to doses of APIs well above the intended dosage. If such exposures occurred frequently and over a substantial period of time, the unintended health consequences to the employees could be severe, even life altering or life threatening. Taking the initiative to combat the risk of overexposure to employees, the parent pharmaceutical companies have turned to the field of industrial hygiene to keep their employees as safe as possible.

1.1.1 MODERN INDUSTRIAL HYGIENE IN THE PHARMACEUTICAL INDUSTRY

Industrial hygiene is the field of occupational safety devoted to keeping workers safe within the workplace. Those who are employed as industrial hygienists strive to protect workers in all industries from chemical, physical, biological, thermal, radiological, and ergonomic agents. A thorough review of the history of industrial hygiene is beyond the scope of this book, but the roots of industrial hygiene can be traced back to the Italian physician Bernardino Ramazzini. He was one of the first physicians to note that certain occupations led to higher incidence rates of specific diseases. His seminal publication "De Morbis Artificum Diatriba" (*Diseases of Workers*) helped pave the way for modern occupational medicine.[1] In turn, this fueled the need to research and evaluate the causative disease agents within the workplaces, thus giving prominence to the field of industrial hygiene (Figure 1.4).

As time progressed, a few significant advancements were made. However, Dr. Alice Hamilton noted in 1910 that occupational exposures to airborne contaminants such as lead and silica were giving rise to obvious diseases in worker populations. She became a pioneer in the developing field of industrial disease, and her work has been regarded as the beginning of the science and practice of industrial hygiene within the United States. Undoubtedly, without the work of Dr. Hamilton, the field of industrial hygiene would be years behind where we are today; however, the actual practice and tenants of industrial hygiene were not yet cemented.

FIGURE 1.4 Pictures of Bernardino Ramazzini (left) and Alice Hamilton (right).

The field of industrial hygiene became much more structured as the 20th century progressed. With the founding of the American Conference of Governmental Industrial Hygienists (ACGIH) in 1938 and the American Industrial Hygiene Association (AIHA) in 1939, there were two authoritative bodies to give credence to the field as well as provide direction for future growth. These organizations began by formally defining the field of industrial hygiene as the science and art devoted to the anticipation, recognition, evaluation, and control of workplace hazards with the intent of keeping employees safe while at their place of occupation. This formal definition of the field has evolved over time as industrial hygienists have continued to expand and refine their craft. The current definition of industrial hygiene is the anticipation, recognition, evaluation, control, and communication of workplace hazards. The aspect of communication was recently added after the realization that even with all the work that an industrial hygienist may perform, without proper communication to the workers and management, there will be no understanding of the hazards present or the solutions implemented.

Throughout the 20th century, the role of the industrial hygienist was to evaluate workplaces by sampling for known or suspected harmful agents and compare the results to acceptable levels. If the acquired samples were satisfactory, the site was considered to be in compliance and nothing further occurred (with the notable exception of the required communication of results to employees); however, if the results were out of compliance, the industrial hygienist would recommend a variety of ways to reduce employee exposures. These methods, referred to as controls, consist of engineering and administrative controls as well as the use of personal protective equipment. The role and value of the industrial hygienist were not held in high regard, perhaps as a necessary evil as far as management was concerned, and far too often, the recommendations given were not properly implemented.

This approach, however, has significant flaws. First, the regulatory levels mentioned above refer to the Occupational Safety and Health Administration (OSHA) permissible exposure levels (PELs) published in 29 CFR 1910.1000. These are the legally enforceable airborne levels of a variety of chemicals in the United States. Unfortunately, these levels are for only a small fraction of all chemicals used in industry, and there are far more substances without PELs associated with them than those with assigned values. Additionally, the PELs on record were adopted from the 1968 ACGIH threshold limit values (TLVs). The TLV list is a recommendation made by scientists who continually update the list by examining new data. As with all scientific fields, as new information is obtained, the allowable levels to chemicals are refined. Unfortunately, such refinement has rarely taken place with the OSHA PELs. The enforceable levels may not reflect current scientific thinking and public health opinion. Consequently, businesses and industries which choose to simply enforce the legally required exposure levels are not acting in the best interest of their employees.

In line with this logic, the pharmaceutical industry realized that many of the substances utilized were specifically designed to have physiological and biological effects on humans at low doses (a theme which will be revisited many times throughout this book). Moreover, the substances being used (active drugs) did not have any regulatory airborne levels associated with them. The industry as a whole took a proactive stance and decided that the companies themselves would essentially self-regulate the exposures of their employees in order to protect their greatest assets. But without any regulatory yardstick against which to measure (e.g., PELs), the pharmaceutical industrial hygienist was not left with much to work. To work around this shortcoming, the pharmaceutical industrial hygienists turned to the budding field of risk management.

1.1.2 RISK MANAGEMENT OVERVIEW

The rationale for controlling risks within an organization may seem straightforward, but curiously there is often not a clearly defined purpose or reason for implementing a risk management plan. Risks should be managed as a means to protect aspects of an organization which create value. In the past, value was attributed to those resources which directly generated revenue, such as machinery or a building's infrastructure. However, a distinct paradigm shift has occurred within businesses over the past 20 years, and now organizations have realized intrinsic value in a myriad of ways. These have grown to include concepts such as supply chains (products cannot be produced without raw materials), product quality (nobody wants a poorly made product), and perhaps most important of all, people. The employees of an organization are the driving force of the culture toward achieving the mission and vision of the organization as well as the manufacturing, marketing, and distribution of its products. Managing the risks to employees is, therefore, one of the most important and fundamental components of a risk management program.

Risk management can be defined as "coordinated activities to direct and control an organization with regard to risk".[2] In turn, a risk can be thought of as an undesired event or occurrence which can impact an organization. Consequently, risk management is a series of activities which are intended to direct and control undesired events from occurring, thus preventing a typically negative outcome or minimizing its impact. An alternative and usually more cited definition of risk is the result of an identifiable hazard and its likelihood, or probability, of occurring (we will refer to this particular definition quite a bit). There are innumerable hazards present in all organizations, but they do not necessarily pose a significant risk if they are not likely to occur. While this makes intuitive sense, most hazards would not be identified and properly controlled without appropriate risk management. The field of risk management became standardized with the publication of ISO 31000. Devoted entirely to the adaptation and inclusion of risk management activities into all facets of an organization, the standard established a clear path forward for organizations to begin proactively assessing and mitigating all types of risks.

Building on the previously stated definitions, the ISO standard outlines a framework for organizations to systematically identify risks and develop solutions. The process flow is shown in Figure 1.5. Prior to managing any sort of risk, it is imperative to establish the scope, context, and criteria for the identified risks. These three concepts lay the bedrock upon which the entire risk assessment and subsequent management is based. Without these clearly articulated and defined, the risk management process is simply an academic exercise in futility which is devoid of direction and purpose. It is not exaggeration to say that the first three concepts are unquestionably the most important and require the most input and time devoted to them.

The scope of a risk management process asks the basic questions that the process is intended to answer. This is particularly important because the risk management standard is written in such a way

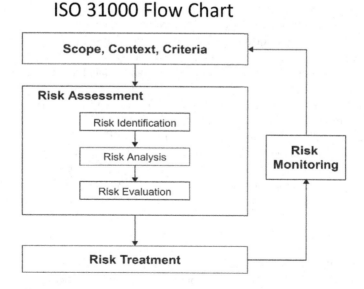

FIGURE 1.5 The general ISO risk management workflow.

that it can be applied to any aspect of an organization. Thus, it can be applied to a specific project, a single portion or piece of equipment on a process, or an entire program. Stating outright what the plan is intended for allows the operators to ask the appropriate questions and utilize the appropriate tools. A common goal of the scope is to define the objectives of the assessment and the expectations. In turn, this will provide a much clearer picture of the resources required to facilitate the assessments and establish realistic time frames for achieving the outcomes. A final advantage of documenting the scope is that it limits "scope creep". As mentioned, the risk management process applies to virtually anything, and no process or equipment exists in a silo. A common phenomenon when performing risk management is to stray from your intended purpose, inadvertently including ancillary processes or related projects to your desired goal. Explicitly documenting the purpose of the risk management process will significantly limit the amount of scope creep that occurs as you progress in an evaluation.

In a similar vein, the context of the risk management plan should also be clearly laid out prior to beginning the process. If the scope of the process attempts to answer what users are aiming to achieve, the context is the answer to why the risk needs to be managed. Risks and their associated management impact the mission and vision of the organization. On some level, whether it be that of top management or on the shop floor, unmanaged risks serve to impede the organizational progress toward fulfilling the desired mission and vision. The impact of this impediment can be any number of scenarios, from lost production time, property destruction, employee turnover, or even damage to the organization's reputation. The context of a risk management plan, therefore, spells out the need for the plan and how it helps the organization achieve its goals and move forward, especially regarding external and internal stakeholders, such as employees.

Rounding out the pre-work activities is establishing the criteria for the risk management plan. This is perhaps the most difficult concept to develop and even more difficult to gain acceptance and buy-in from management for implementation. The criteria lay the groundwork for all decisions which result from a risk assessment protocol, so the importance of determining the criteria for what constitutes a risk cannot be overstated. Compounding this criticality is the fact that all parties involved bring their own biases, skewed perceptions, ulterior motives, and personal beliefs to the discussion when outlining the parameters for criteria. And since risk management teams are a multi-disciplinary group comprised of individuals from a range of backgrounds, there are likely to be disagreements during this stage.

Before any risks can be managed, however, it is necessary to know if a risk exists. Within the framework of ISO 31000 lies the risk assessment, a concerted process used worldwide to identify and label risks. The process itself is broken down into individual steps by the standard, including risk identification, risk analysis, and risk evaluation. Numerous techniques exist to carry out a risk assessment – in fact, the ISO put forth an additional standard detailing a variety of risk assessment methods and how they can be used most beneficially for the process.[3]

The ISO 31000 risk identification is exactly how it sounds: labeling and describing risks that may be present. At this stage of the assessment process, likelihood of occurrence has not been evaluated, and thus an actual risk cannot be assigned; rather, the hazards are identified, labeled, and prioritized. There are numerous tools and techniques available to the individual performing the risk identification.[4] There are numerous ways to recognize if a hazard exists from a particular chemical or material, as we will see later. These sorts of analyses are what are typically considered when evaluating workplace exposures. Regarding other forms of hazard identification, some other methods to identify a hazard include preliminary hazard analysis (PHA), failure mode and effects analysis (FMEA) and fault tree analysis (FTA). These tools are often utilized before a process is executed for the first time (such as when a piece of equipment has been installed but not used) and are very useful for simply identifying hazards. It is imperative that a diverse team be assembled to assess the hazards as a singular person is likely to view the hazard spectrum through a biased lens.

The next step of the assessment process is the risk analysis. According to the standard, the risk analysis is the stage which considers the consequences (outcome), scenarios, likelihood, controls, and their effectiveness.[5] Risk analysis is the often subjective assessment of the risk team to determine how likely the identified risk is to pose a threat. In other words, the risk analysis attempts to predict the probability of the risk occurring. This aspect of the risk assessment is usually qualitative in nature, rating the likelihood of occurrence with terms such as "rarely", "occasional", or "frequent", although the field of quantitative risk assessment is becoming more commonplace.[6] The risk analysis step is one which may be most familiar to an industrial hygienist, as he or she routinely conducts these sorts of assessments on the regular. Indeed, more on this particular point will be expounded upon in Chapter 4.

The final step in the risk assessment process is the risk evaluation. Essentially, the evaluation is the final endpoint at which the team decides if the company must act to reduce the risks they have analyzed. If the assets in question are subjected to sufficient stressors that a significant risk exists, then decisions must be made. Such decisions can be pre-determined in the Scope, Context, and Criteria phase depending on the outcome of the analysis and how the risks are scored, or the results can be used to generate a lively and productive discussion on how best to proceed. In general, the options are: no action taken (the company can absorb or tolerate the current amount of risk for the asset), to transfer the risk (a method not typically taken by companies outside of the insurance realm), or to reduce the risk (actions taken to mitigate the hazard or its likelihood of occurrence). A fourth option, avoidance, is also available, but this is not typically utilized in the pharmaceutical sector.

A major caveat to this is that risk assessments from an industrial hygiene perspective involve substances that can have a harmful effect on employee health. Performing a risk assessment on a process or material which can affect the health and well-being of people is vastly different than assessing whether a company has other risks, such as significant cybersecurity vulnerabilities. Such a difference has long been recognized by risk professionals and public health professionals alike. Indeed, in 1983, the National Research Council (NRC) issued a landmark publication entitled "Risk Assessment in the Federal Government: Managing the Process" in which the process of conducting a human health risk assessment was clearly laid out as a series of steps consisting of hazard identification, dose-response assessment, exposure assessment, and risk characterization (Figure 1.6).[7] Since the formalization by the NRC, this particular method of risk assessment has been widely adopted as the industry standard by groups such as the US Environmental Protection Agency (EPA),[8] the World Health Organization,[9] and the American Chemical Society. Indeed, most safety risk professionals are adequately acquainted with the NRC risk assessment paradigm. The risk assessment methodology put forth within this book is based entirely off the NRC recommended approach.

Once a risk treatment strategy has been realized and implemented, the risk management process is still not complete. ISO 31000 requires a continuous monitoring and evaluation protocol for all managed risks to ensure that the process has indeed reduced the perceived risks. For example, if access to a building was deemed a risk and the treatment for this was to install Radio frequency

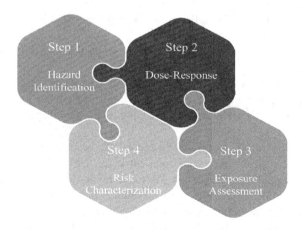

FIGURE 1.6 The National Research Council (NRC) human health risk assessment process.

identification (RFID) key tags to limit access, the company should not simply install the door locks and consider it a job well done. Rather, a continuous process of ensuring the system is functioning properly is put in place. The system may be randomly audited to see if it can be bypassed or if the system recognizes other RFID tags from other systems. A continuous monitoring strategy for any risk is vital to ensuring that the methods used to manage the risks are effective. In Figure 1.5, this essentially starts the entire risk management over again, forcing a re-evaluation of the risk. Such a practice is common within the risk management community and periodic review of all managed risks ensures that systems are working as intended.

Throughout the process, ISO 31000 stresses that all decisions and criteria should be not only documented for future reference but also communicated to all stakeholders. More often than not, these are simply "email blasts" from a communication office or a simple newsletter that is placed in a normally populated location. Maintaining reports and lines of communication with all parties involved is important to ensure that everyone is working to minimize the identified risks in the same manner.

This brief overview of risk management demonstrates a very general approach to managing risk which can be adopted by any industry. In fact, most organizations have performed one or more of these tasks, but perhaps not all of them together in a formal fashion. As companies have begun to focus their approach from a compliance-based focus to a risk-based focus in order to protect a wider array of assets, the risk management philosophy pioneered in ISO 31000 has become more widely adopted.

1.2 INDUSTRIAL HYGIENE AND RISK MANAGEMENT WITHIN THE PHARMACEUTICAL INDUSTRY

Integrating risk management into the pharmaceutical sector did not happen overnight, nor did it initially happen with industrial hygiene. Perhaps the first real implementation of risk management, and certainly the most notable, was within the realm of quality. In August of 2002, the US Food and Drug Administration (FDA) released their new initiative for the pharmaceutical sector. Entitled "Pharmaceutical Current Good Manufacturing Practices (cGMPs) for the 21st Century", the goal was to modernize the regulations relating to pharmaceutical manufacturing while ensuring product quality.[10] The final report was issued 2 years later and listed six key topics which the FDA asked the pharmaceutical industry to embrace:

- Encourage the early adoption of new technological advances by the pharmaceutical industry;
- Facilitate industry application of modern quality management techniques, including implementation of quality system approaches, to all aspects of pharmaceutical production and quality assurance;
- Encourage implementation of risk-based approaches that focus both industry and agency attention on critical areas;
- Ensure that regulatory review, compliance, and inspection policies are state-of-the-art pharmaceutical science;
- Enhance the consistency and coordination of FDA's drug quality regulatory programs, in part, by further integrating enhanced quality systems approaches into the agency's business processes and regulatory policies concerning review and inspection activities.

Of the six points highlighted within the FDA's report, the second and third points were widely pushed into the pharmaceutical sector with apparent vigor. Only a year after the publication of the report, the International Conference on Harmonization (ICH) released *Q9: Quality Risk Management*, a guidance document for the pharmaceutical manufacturing sector.[11] The document outlined a scalable plan for the adoption and implementation of quality risk management (QRM) into all aspects of pharmaceutical production, including quality assurance and manufacturing. A year later, the FDA issued its own QRM guidance document which adopted and aligned with the recommendations of the ICH.[12]

ICH Q9 presented several protocols and tools for identifying risk to quality-based systems. A comparison of the QRM protocol to ISO 31000 reveals that ICH modeled their process upon the international standard. In fact, it is almost identical, with only minor additions for clarity. This similarity is by no means an accident or a coincidence. ISO 31000 operates essentially as a performance-based standard. In this regard, the standard outlines the goals and direction for achieving appropriate risk management but does not specifically spell out *how* to achieve it. This absence of information is intentional as it was realized that there is no one-size-fits-all approach to risk management. Organizations are permitted to customize the execution of their risk management program so long as they follow the general process outlined in the standard. By adopting the ISO 31000 process, the ICH (and by extension the FDA) granted much-needed flexibility to the pharmaceutical industry for the implementation of their QRM programs (Figure 1.7).

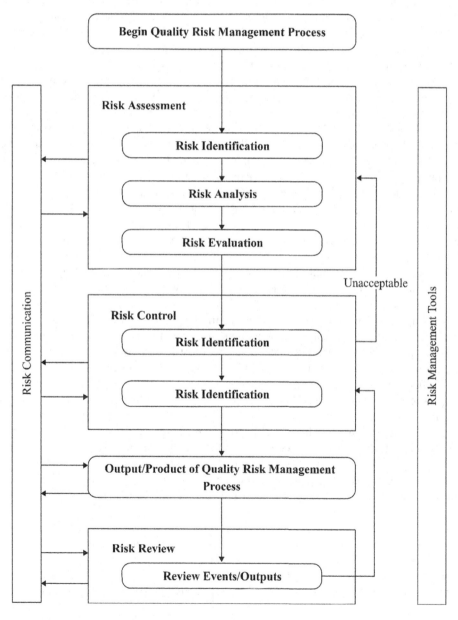

FIGURE 1.7 The ICH Q9 QRM methodology.

Pharmaceutical companies and their subsidiaries did not take long to embrace the QRM guidelines. In short order, committees were being established to assess the quality risk for every aspect of a product lifecycle and risks were being cataloged within risk registers. Over the last 15 years, the process has been refined as organizations learned to cope and deal with risks as well as how to better evaluate them. Individuals with professional quality risk certifications are becoming more and more commonly employed to continually improve and lead QRM programs. Whole books on the subject for the industry have been published which serve as references for the industry.[13] Quality-specific audits are now routine among those in the pharmaceutical industry. QRM is now completely integrated into the industry and has thus far paid dividends.

The reader may recall that earlier we stated risks should be managed to protect assets which create or have value. Without question, ensuring appropriate product quality is a critical component of the pharmaceutical industry. After all, how else would we be able to ensure the patient or consumer that the product they are taking is manufactured with the highest degree of purity? There is certainly value in such an endeavor. But we also stated that the most valuable asset that any organization has is its employees. This is where industrial hygiene begins to play a vital role to companies, especially pharmaceutical and consumer healthcare organizations.

Within the pharmaceutical sector, the industrial hygienist does not play a passive role of simply determining if exposures comply with published standards; rather, the hygienist plays one of the most dynamic roles within a safety group. Gone are the days in which the role of the hygienist was simply to sample and collect data. In the modern era within the pharmaceutical sector, the role of the industrial hygienist is one of the most all-encompassing. They are expected to be not only knowledgeable and fluent in the language of traditional industrial hygiene, but also knowledgeable in the fields of data analysis, particulate sampling, air flow, ventilation, containment, project management, and perhaps most important of all, communication. More often than not, the site industrial hygienist will be tasked with ascertaining appropriate air flow for an operation, troubleshoot ventilation systems connected to local exhaust ventilation (LEV), evaluate various containment options for a specific set of substances in a particular occupational exposure band (OEB), or weigh in on designs for a completely new process altogether. Moreover, many retrofit projects may fall under the purview of the site hygienist, making them not only the expert in assessing solutions, but also overseeing the entire project itself. Truly the role of the pharmaceutical Industrial hygienist (IH) has become a dynamic and indispensable role within pharmaceutical sites.

Another important aspect to consider is that within the pharmaceutical field, industrial hygiene does not exist in a silo; instead, it exists at a unique junction of three seemingly distinct and unrelated fields: safety, engineering, and manufacturing. Most readers will understand each field on its own, and perhaps even where two may overlap. Where safety and engineering come together, they form effective engineering controls which serve to enhance the workplace and make it safer. Similarly, where engineering and manufacturing overlap is the implementation of newer, more effective machinery to enhance the production of materials. It is at the unique junction of all three disciplines that industrial hygiene has the distinction of residing. The hygienist is expected to utilize science and tools of the trade to keep employees safe (safety) by implementing and/or maintaining additional controls (engineering) in such a way that does not significantly impede the bottom line (manufacturing). Attempting to balance and satisfy stakeholders from all three disciplines is often a daunting task. A fourth group, quality, is always present and must be taken into account at all stages of change implementation (Figure 1.8).

In order to accomplish this daunting set of responsibilities while ensuring employees are properly protected, pharmaceutical companies have integrated risk-based approaches into their industrial hygiene programs. The following chapters of this book detail the modern approach taken by most of the industry to incorporate ISO 31000 principles into their industrial hygiene programs to better protect their personnel. Of note is that this book will focus solely on the protection of employees

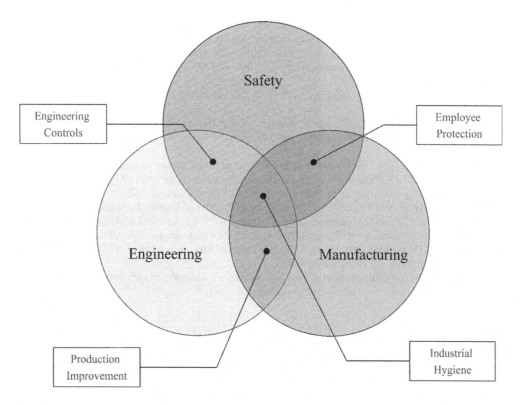

FIGURE 1.8 Pharmaceutical industrial hygiene occupies a niche role at the intersections of safety, engineering, and manufacturing.

from airborne chemical hazards; however, industrial hygiene encompasses protecting employees from all sources of hazards, including biological and physical hazards. While the same risk management principles can be applied to the control of these hazards as well, we will be focusing solely on airborne chemical hazards in order to focus our content.

1.3 SUMMARY

In this chapter, we explored the critical role of pharmaceutical companies in our everyday lives and the significant demand for their products. In order to meet this ever-increasing demand, pharmaceutical companies are now significantly investing in their employees, which are now considered to be their most important asset. The best way to protect employees from occupational hazards is to utilize a risk management approach. Pharmaceutical companies have integrated ISO 31000, the risk management standard, into virtually all facets of their organizations. This was first done within the quality sector and was soon integrated into all other aspects, including industrial hygiene.

NOTES

1 Ramzzzini, B. *Diseases of Workers*. New York, Hafner Publishing Co., 1964.
2 International Organization for Standardization. (2018). *Risk Management*. (ISO Standard 31000:2018).
3 International Organization for Standardization. (2019). *Risk Management Risk Assessment Techniques*. (ISO Standard 31010:2019).

4 Ostrom, L. and Wilhelmsen, C. (Eds.). (2019). *Risk Assessment Tools, Techniques, and Their Applications.* (2nd digital ed.). John Wiley & Sons. doi: 10.1002/9781119483465.

5 International Organization for Standardization. (2018). *Risk Management.* (ISO Standard 31000:2018).

6 International Organization for Standardization. (2019). *Risk Management Risk Assessment Techniques.* (ISO Standard 31010:2019).

7 NRC. *Risk Assessment in the Federal Government: Managing the Process.* Washington, DC: National Academy Press; 1983.

8 Risk Assessment Forum. (2014, April). *Framework for Human Health Risk Assessment to Inform Decision Making.* (EPA/100/R-14/001). Environmental Protection Agency. Office of the Science Advisor.

9 WHO human health risk assessment toolkit: chemical hazards, second edition. Geneva: World Health Organization; 2021 (ICPS harmonization project document, no. 8). License: CC-BY-NC-SA 3.0 IGA.

10 US Food and Drug Administration. (2004, September). *Pharmaceutical CGMPs for the 21st Century - A Risk-Based Approach. Final Report.* Department of Health and Human Services.

11 ICH Harmonized Tripartite Guideline. (2005, November). *Q9: Quality Risk Management.* International Council for Harmonizations of Technical Requirement for Pharmaceuticals for Human Use.

12 US Food and Drug Administration. (2006). *Q9 Quality Risk Management. Guidance for Industry.* Department of Health and Human Services.

13 Mollah, A. H., Long, M., Baseman, H. S. (Eds.). (2013). *Risk Management Applications in Pharmaceutical and Biopharmaceutical Manufacturing.* John Wiley & Sons.

2 Risk Assessment
Hazard Identification and Dose-Response Evaluation

2.1 INTRODUCTION

The manufacture of products which contain active pharmaceutical ingredients (APIs) is an inherently risky operation. In fact, ICH Q9 clearly states "…manufacturing and use of a drug (medicinal) product, including its components, necessarily entail some degree of risk". It is not possible, nor should it be practical, to have zero risk within a good manufacturing practices (GMP) facility which handles, packages, or manufactures APIs. Rather than attempt to achieve zero risk to a site, material, or to employees, it is far more reasonable to effectively manage risks. ISO 31000 provides a convenient roadmap for all industries to ensure that the nuts and bolts of a risk management program are in place.

As we mentioned in the previous chapter, a risk must first be identified and assessed at all before it can be managed. The risk assessment is a core tenant of the ISO 31000 roadmap and is probably the singular process that most people are aware of. But a proper risk assessment is not a lackadaisical activity that can be carried out on a whim. It is a regimented and structured process for which most assessors receive at least some training. Within the context of the ISO standard, a risk assessment consists of risk identification, risk analysis, and risk evaluation. These steps were enumerated

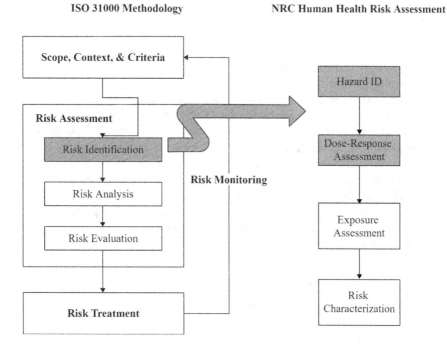

FIGURE 2.1 This chapter will focus on the hazard identification and dose-response steps of the industrial hygiene risk assessment.

DOI: 10.1201/9781003273455-2

in greater detail in Chapter 1. While the standard details what these steps are and what is to be discovered and documented, it does not spell out how to do a risk assessment. There are many ways to carry out a risk assessment, but by and large these techniques are for general risks.

But a human health risk assessment is inherently very different from other risk assessments, and these are of paramount importance to the field of industrial hygiene. Human health risk assessments follow the paradigm suggested by the National Research Council's 1983 landmark publication "Risk Assessment in the Federal Government: Managing the Process".[1] The risk assessment as outlined by the NRC consists of hazard identification, dose-response assessment, exposure assessment, and risk characterization. Since its introduction, this methodology has been widely adopted by many regulatory agencies in the United States and internationally. Importantly, it is also the process employed by industrial hygienists in the pharmaceutical industry.

The reader may observe that the NRC definition of risk assessment does not quite align with the ISO 31000 definition. This modest discrepancy has not gone unnoticed, and efforts have been made to equate the two, thereby allowing the NRC methodology to be incorporated into the ISO framework.[2] The exact terminology utilized by an assessor does not really matter; what matters is the output, the risk in question being identified and characterized.

In theory, a human health risk assessment is a terrific academic and intellectual exercise which allows risk assessors to perform a deep dive into how substances can affect a given population. In practice, however, there are numerous pitfalls due to incomplete data, erroneous data, or simply a dearth of data. Combined with the requirement for a significant amount of expertise to perform one or more parts of the NRC risk assessment, the process is often very frustrating indeed. Furthermore, since the process does vary from traditional risk assessments, the output of a human health risk assessment does not always lend itself to prioritization among risk managers.

Historically, human health risk assessments from an industrial hygiene perspective were conducted in a qualitative manner; that is, they were conducted by visually observing a process and assigning the hazard and likelihood variables by subjective terms. Such practices lead to inconsistent risk assessments and disagreements between assessors, even when defining criteria are established. As we will see later, humans are particularly bad at subjectively assessing exposures based on observation, thus introducing a significant amount of error and bias into the assessment and thereby incorrectly categorizing risks. The pharmaceutical industry has taken significant steps to remove subjectivity by phasing out qualitative assessments and instead using objective quantitative assessments using modern statistical tools.

This chapter will detail how the field of pharmaceutical industrial hygiene has approached the implementation of the NRC risk assessment system and incorporated it into ISO 31000. In addition to describing how the system has historically been used, flaws and shortcomings of the system will also be described and how the pharmaceutical industry has largely overcome these flaws in the risk assessment methodology. Specifically, this chapter will focus on the first two steps of the human health risk assessment: hazard identification and dose-response assessment. Note that any specific criteria mentioned may not be the exact criteria that an organization may use. This chapter is not intended to be a playbook by which every organization should follow; rather, it is intended to be a general guide for readers to establish their own criteria in their risk management programs. The stakeholders within each organization must decide exactly how much risk they can tolerate and what the criteria are for those pre-established limits.

2.2   HAZARD IDENTIFICATION

The first step in every human health risk assessment is the hazard identification. This step sets the tone for the entire industrial hygiene process and attempts to answer the question of whether or not a material poses some sort of health hazard. This is perhaps the most easily recognized step of the risk assessment process, as virtually every person has performed a variation of this step at some point in their lives. As an example, it is not uncommon for a consumer looking to utilize a new herbicide on his or her garden to wonder, and then seek and answer to, if the chemical under consideration is in fact harmful or not. Whether the consumer is aware of it or not, this is in fact hazard identification, the first step of a human health risk assessment.

The NRC defined the hazard identification as the step in attempting to determine whether exposure to a substance would impart an adverse health effect.[1] At the time of the NRC publication, the causation of cancer was of preeminent concern; however, the list of adverse health effects for which a substance is assessed has increased significantly in the decades since the original publication. Some of the more common classifications of chemical health effects include carcinogenesis, developmental toxicity (teratogenicity), reproductive toxicity, immunotoxicity, neurotoxicity, hepatoxicity, nephrotoxicity, cardiotoxicity, pulmonary toxicity, ocular toxicity, dermal toxicity, and mutagenicity.[3] Yet the simple identification of whether a substance exhibits these health effects is not the sole purpose of the hazard identification step. That is, it is not a simple "yes/no" answer. One of the most important, and often overlooked, steps of the hazard identification is the evaluation of the quality of the data. This often begins by evaluating the type of data and its source. The NRC outlined that such information can typically be garnered from any of four sources: epidemiological studies, animal bioassays, in vitro assays, or comparisons of chemical structure (non-test methods) (Figure 2.2).[1]

2.2.1 Epidemiological Studies

For many chemical entities, well-documented epidemiological studies are not common. A typical risk assessor would be lucky indeed to find a well-documented and verified epidemiological study of substances not commonly encountered in the general population. However, the pharmaceutical industry is both blessed and unlucky at the same time in this vein. The FDA requires significant amounts of data to be presented from a host of studies before a substance will be approved for commercial therapeutic use. These not only include traditional toxicology studies but also clinical trials. The clinical trials can be thought of as small-scale epidemiological studies. Phase II and Phase III trials in particular serve as excellent studies for documenting potentially adverse reactions to API dosing. When paired with pre-clinical toxicological evaluations, the new drug submittal process produces a large amount of data from which the industrial hygienist or risk assessor can potentially identify adverse health outcomes.

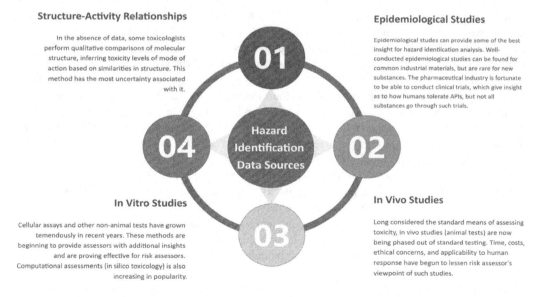

FIGURE 2.2 The four primary means of gathering information for hazard identification are epidemiological studies, animal bioassays, in vitro studies, and structure-activity relationships.

A significant drawback to this route of information gathering is that clinical trials occur very late in the product development life cycle. It is often many years into the project before any trials begin. Prior to this, there are still employees utilizing and handling the material for various operations and tasks which need to be properly protected from harm. Furthermore, not every substance utilized in the pharmaceutical sector undergoes significant evaluation. For instance, numerous non-APIs do not benefit from the luxury of clinical trials. Excipients, flavors and colors, and chemical intermediates do not typically have epidemiological studies for them.

2.2.2 Animal Bioassays (In Vivo Studies)

The second data type for garnering hazard data is from animal studies. These types of analyses are perhaps the most sought during a risk assessment, as they represent a historical preponderance of data. The logic behind animal testing is almost undeniable: a series of doses are administered to an animal test population and compared against a control group to see at which dose an adverse effect appears (or disappears). Perhaps the most notable animal bioassay is the acute toxicity evaluation, also known as the LD_{50} assessment. First introduced in 1927, the LD_{50} assessment of chemicals served as the primary means of evaluating the hazards of a chemical prior to further evaluation.[4] In the years since its introduction, the LD_{50} assessment has undergone extensive refinement. Numerous other animal bioassays have also been developed for different and specific endpoints. In order to ensure results could be compared between experiments, such assays became standardized test methods. Organizations such as the Organization for Economic Cooperation and Development (OECD) have published guidelines for how specific animal bioassays should be conducted and the results reported. For example, OECD test 406 is specifically for the assessment of skin sensitization of a chemical, while test 433 evaluates acute toxicity via inhalation rather than other routes of administration. Essentially every potential adverse health effect previously mentioned has a defined OECD test method associated with it. By following such guidelines, researchers enable their results to not only be reproduced but compared globally, allowing for consistent analyses.

The use of animal models dominated the toxicological world for decades, and the results were considered to be directly applicable to the prediction of human health. In fact, the NRC stated that the "… inference that results from animal experiments are applicable to humans is fundamental to toxicologic research".[1] However, researchers have begun to realize that animal models are not necessarily good or predictive models for how human systems will react to similar chemical exposures. A rather well-known example of this lies in the history of saccharine. Animal testing initially identified the artificial sweetener as a carcinogen after rodents developed bladder tumors. The substance was labeled as a known carcinogen and warning labels followed the substance for decades. It was only after breakthroughs in testing that the biological process which causes the formation of the tumors in rodents was firmly established. Furthermore, the same tumorigenic mechanism cannot occur in humans. Consequently, saccharine is not a carcinogen to humans. After these realizations, the carcinogen rating from the International Agency for Research on Cancer (IARC) was downgraded from 2A (probably carcinogenic to humans) to 3 (not classifiable as a carcinogen to humans). This example demonstrates how animal models can give erroneous results that can cause significant downstream effects, particularly in risk assessments and in regulatory scenarios.

In addition to their rather poor predictive capacity for human health, there is a significant effort to move away from animal testing from an ethical perspective.[5] The current scientific ethics dogma is to embrace the concept of reduction, refinement, and replacement of animals in testing, commonly referred to as the "3 Rs". The concept of reduction applies to the use of the bare minimum number of animals used in an assay, when they need to be used at all. Theoretically, a researcher could use a substantially large number of animals to conduct an assay, but the "reduction" aspect argues that meaningful and useful data can be acquired with a very small test population. Many OECD test protocols have been revised to incorporate this philosophy into their methods. The "refinement" aspect refers to altering the test protocols to mitigate or outright eliminate any unnecessary pain, suffering, or discomfort experienced by the animal during the test. Again, many OECD tests have been revised in an effort to embrace this notion. The final "R" is the "replacement" aspect. In this context, the effort is to replace existing animal test methods with alternatives which do not utilize animals at all. Such an effort is a recall to the previous point that animal models are not necessarily good models for human health responses. The encouragement to replace animal tests with non-animal methods leads to the third data type deemed acceptable by the NRC: in vitro assays.

2.2.3 In Vitro Assays

The field of in vitro toxicology has become an undeniable force within the occupational health realm and has truly revolutionized how scientists go about assessing chemicals. In 2007, the NRC released a report detailing a new vision of how toxicological data would be obtained for new chemicals and environmental agents.[6] Entitled "Toxicity Testing in the 21st Century: A Vision and Strategy", the document called for a complete overhaul of the current toxicity testing regimen. Specifically, the NRC detailed the inefficiencies of current testing strategies utilizing animals; that is, they are expensive, unreliable, and inhumane. The proposed solution called for a significant increase in cell-based work (in vitro studies).

Over the last several decades, the efficacy, sensitivity, reliability, reproducibility, and breadth of in vitro toxicity testing have all increased substantially. Whether in response to the NRC call for increased emphasis on in vitro testing or not, the end result has been the same. Numerous in vitro assays exist for the same endpoints that OECD tests for. As an example, OECD Test 406 utilized guinea pigs to assess for skin sensitization, or the prevalence of a chemical to illicit an allergic reaction after repeated dermal contact. Yet this test protocol now has an in vitro alternative, OECD Test 442D. Toxicity results obtained following this in vitro protocol are considered equivalent to the previous animal-based methodology. Other OECD methods employing in vitro models include tests for phototoxicity, genotoxicity, and skin corrosion to name a few. The field of in vitro toxicology is rapidly expanding and is not just relegated to assays developed or accepted by the OECD.[7] Numerous examples of in vitro assays can be found within the toxicology literature for virtually any type of endpoint. For instance, published assays exist which evaluate specific aspects of cardiotoxicity, reproductive toxicity, and immunotoxicity. From a risk assessment standpoint, such tests provide invaluable information for identifying particular hazards and thereby classifying the chemical appropriately.

While there are numerous published assays which may provide toxicity insight to chemicals, there are two confounding issues surrounding these tests. The first is that just because an assay exists does not mean it has been applied to a particular chemical of interest. For non-standardized assays (i.e., those published in academic journals and not yet adopted by authoritative bodies), it is highly unlikely that a researcher will have applied a specific test to a potentially obscure chemical of interest to a risk assessor. The pharmaceutical industry does have a significant advantage that they have the capability to perform such tests at their convenience, but this may not be the case for smaller start-up companies. Thus, they are left with the same problem of not having sufficient information or data to identify potential hazards. The second confounding point is that just because an assay is developed and published does not necessarily mean it is a widely useable assay or that a researcher executed it properly. This gets into a separate issue in which the assessor must evaluate the quality of the data and the study (more on this later).

The rise in in vitro toxicology studies and assays for hazard identification is not only due to concerns regarding costs, reliability, and ethics, but also for legal reasons. In 2019, the US EPA announced it would ban all funding and studies for toxicity utilizing animal testing.[8] Similarly, in the summer of 2022, the US House of Representatives passed the FDA Modernization Act, a bill that would end the FDA mandate for animal testing with new drug applications. The FDA Modernization Act does not outright ban animal testing, but no longer makes it a requirement for animal testing data to be submitted for new substances. Both pushes from the EPA and Congress indicate a significant paradigm shift in how risk assessors in the pharmaceutical and chemical industries are to acquire requisite information to determine adequate safety levels for not only consumers but also employees within these industries.

Within the last two decades, an additional entry into the alternative to animal testing universe has emerged: in silico testing. In contrast to testing toxicity endpoints on animals (in vivo testing) or cultured cells (in vitro testing), in silico methods utilize computational approaches to model the effects of compounds. This is a budding field which has received an increasing amount of attention.[9] One of the most common uses of in silico toxicology has been the implementation of structural alerts by which a software program analyzes a chemical structure and informs the assessor to a potentially problematic functional group or moiety.[10] Frequently such alerts are for carcinogenic groups such as nitrosamines, aromatic amines, or dioxin-containing compounds. The structural alert does not give any additional insight, such as a potential dose or concentration at which the effect may begin to manifest itself in a human population. Other in silico methods have not been as restrictive to simple identification. Methods have been developed to assist risk assessors in identifying potential endpoints for acute toxicity and mutagenicity. Such models exhibit a significant amount of uncertainty, as would be expected, since they are merely models and contain built-in assumptions. However, with further refinement this exciting field would serve as an excellent means to rapidly assess occupational chemicals. A recent example of using in silico toxicology to identify skin sensitizers showcased how such techniques can be applied, and the models further refined and calibrated to produce more accurate results.[11] The use of in silico methods could serve to potentially identify occupational exposure limits for chemicals, especially those in the pharmaceutical field. The computational application to intermediates, by products, and even final APIs would greatly facilitate the field of risk assessment; however, this has not become a mainstream practice as of yet.

2.2.4 STRUCTURE COMPARISON

The fourth class of data for hazard identification as identified by the NRC is structural comparison. We have mentioned that relatively few substances have had a complete analysis of toxicological effects, but those that have enjoy an array of structural diversity. Many of these substances range from simple solvents to complex structures of anthropogenic origin, including pharmaceuticals. For the significant number of new or simply unassessed substances, it is possible to compare molecular structures to those which have copious amounts of available data. The logic is that compounds of similar structure would theoretically exhibit similar properties such as solubility, ADME characteristics, pH, and toxicity.

This approach has formally been designated as a "read-across" methodology. Read across, or RA, follows an accepted protocol for establishing plausible health limits. The substance being evaluated is assigned as the "target molecule". Once the target is established, the risk assessor then decides if the RA will be performed against a single substance (analog approach) or against multiple substances of similar structure (category approach).[12] After the target and the subsequent approach is decided upon, the comparison substances must be chosen and assessed for appropriate similarities. This begins with an assessment of structure. Once done, assessment of potential hazards by comparison of the target molecule to similarly structured ones becomes a relatively quicker exercise than any of the other previously mentioned hazard identification techniques.

The use of RA can be used to fill many toxicological gaps in the risk assessment process, not just hazard identification. For instance, RA has been used to estimate points of departure or dose levels for various endpoints such as chronic toxicity and mutagenicity. Naturally, for a technique that is deriving such values from structural comparison, there is an inherent amount of uncertainty associated to the final outcomes.[13] Uncertainty exists even with the previously discussed data types, but there is generally more uncertainty surrounding RA outcomes, given that they are based on comparison and not tests on the target compound. Such uncertainties must be espoused and explained by the assessor in the report, which can then be either accepted or rejected from use in the final assessment.

Although identified by the NRC as a lower quality or less preferred data type, structure comparison has seen an increase in popularity in recent years. The cause for this is obvious, as the number of substances entering the workforce and/or commerce is ever increasing. It is simply impossible from a cost and time perspective to be able to fully test every new molecule, intermediate, and byproduct for toxicity. The increase in RA popularity and the demand for at least a first-tier assessment of substances led to a formalized description of the read-across assessment framework (RAAF) by the European Chemicals Agency.[14] From an industrial hygiene perspective in the pharmaceutical world, techniques such as RA will continue to gain prevalence as new molecular entities begin to be prepared and the need for rapid assessment and categorization of substances by hazard becomes paramount. For identifying hazards in the risk assessment process, proficiency in techniques such as RA will undoubtedly become more commonplace in subsequent years.

2.2.5 FINDING THE RELEVANT HAZARD IDENTIFICATION DATA

The previous discussion sections detailed the types of data which a risk assessor should use or be familiar with in making informed decisions for identifying health hazards from chemical substances, including APIs. But knowing the types of data and what to look for is one piece of the puzzle. Another significant obstacle is knowing where to look for the hazard identification data.

Within most organizations and industries, the most common location a risk assessor begins his or her search for hazard information is typically the safety data sheet, or SDS. A legal requirement under the OSHA Hazard Communication Standard (29 CFR 1910.1200), an SDS is intended to be a concise document which relays all the relevant hazard information for a specific chemical. Each SDS is presented in a standardized 16-section format and is supposed to state all physical hazards, health hazards, emergency response information, proper storage recommendations, physical properties, exposure thresholds, and toxicological effects of exposure. Furthermore, the identification and classification of hazards are intended to align with the United Nations Globally Harmonized System of Classification and Labelling of Compounds (GHS) so that all SDSs will present the same information, regardless of the source.[15] On the surface, such documents would be a welcome and vital resource for chemical handlers and risk assessors alike.

However, the reality of the utility of SDSs is far more somber. First, there is no requirement that an SDS must state where the document author obtained his or her information. References are typically not provided, so the validity of the information must be taken on faith. For any listed relevant health hazards (such as respiratory sensitization, for example), a pictogram is presented along with the relevant GHS hazard statement (H-statement), and perhaps some toxicity information is presented later in the document. There is no citation telling the reader where the source of said information was found. For seasoned risk assessors, the lack of transparency and the inability to assess the quality of the study giving rise to the hazard statement is often a source of frustration. Second, and perhaps more importantly, the quality of information between different SDSs is often lacking. For instance, it is perfectly acceptable for one SDS to state "no data available" for a particular field, whereas an SDS for the same substance from a different distributor will have a wealth of information. The inconsistency of SDS authoring is another significant source of consternation among risk assessors. The "no data available" statement is particularly common for new APIs being shipped between sites or for reagents. Another example of the lack of quality of many SDSs is the evident "copy and paste" from one document to another. A well-known example of this can be found within the SDS for water. All too often, in the Emergency Response Section, the document instructs the reader that if water gets into the user's eye, then they should immediately flush their eyes with water for 15 minutes. The obvious "copy and paste" effort questions the credibility of the SDS from the manufacturer. The absence of overall helpfulness of an SDS has been noted by chemists.[16]

Yet despite these shortcomings, many human health risk assessment invariably begins with the SDS for a substance. Their easy accessibility and readability lend themselves to a quick assessment for hazard identification. Many APIs prepared by pharmaceutical companies will thoroughly test their materials and can adequately document the associated hazards in accordance with the GHS system. However, early-phase drug candidates and other chemical substances may not have sufficient testing performed, leading to the ubiquitous "no data available" phrase mentioned previously. All too often a risk assessor interprets a lack of hazard data to mean "no hazard present". This single commonly encountered issue in risk assessment is enough to discourage assessors from beginning with SDSs. Simply because a SDS does not identify any hazards is not the same thing as the substance being non-hazardous.

Rather than starting with SDSs, seasoned risk assessors in the pharmaceutical sector turn to alternative sources of information. Primary among these are publicly available databases which provide a convenient search feature for substances. Examples of such useful databases include PubChem (www.pubchem.ncbi.nlm.nih.gov), the U.S. National Toxicology Program (NTP, www. ntp.niehs.nih.gov), the U.S. EPA Integrated Risk Information System (IRIS, www.epa.gov/iris), the European Chemicals Agency (ECHA, www.echa.europa.eu/home), and the OECD eChemPortal (www.echemportal.org/echemportal/substance-search). These databases and others provide summaries pertaining to health hazards, physical hazards, and molecular properties, among others. The summaries are pulled directly from primary literature sources. Importantly, the references and sources of information are always given, allowing the risk assessor to go back and read the original study to assess its validity. This aspect of the hazard identification and data review process is crucial and is sorely lacking if only SDSs are used as the source materials.

While providing significantly better information, performing database searches on substances is extremely time consuming and tedious, especially if the assessor is going back and checking the validity of source materials. Additionally, it is a very common occurrence for an assessor to search for a new substance, an intermediate, or an obscure reagent only to find a dearth of information in the databases. These two drawbacks form the chief issue in chemical risk assessment and human health risk assessment: there simply is not very much information on the vast majority of substances in commercial use. In recognition of this significant limitation, international efforts have been made to facilitate this initial component to the risk assessment process (after all, if no hazards can be identified, then a risk assessment cannot be performed).

2.2.6 Hazard Identification Summary

The identification of health hazards from chemicals is the first step in a human health risk assessment as outlined by the NRC. This crucial step, which is all too frequently fraught with uncertainty, sets the foundation for the entire remaining risk assessment. This is because some health hazards are considered far more serious and dangerous than others and are given additional weight when considering setting occupational exposure limits (OELs) or designing controls. Whether the hazards are identified through epidemiological studies (i.e., clinical trials), animal studies, in vitro studies, or structural comparisons, each method has its own limitations and uncertainties. The use of modern technology by way of databases to search through the vast array of literature to identify health hazards has greatly facilitated hazard identification process, yet improvements are still needed.

A final note on hazard identification regarding the risk assessor. Very often the individual conducting the hazard identification is the site industrial hygienist. Although the process is vital to the role of the industrial hygienist, he or she may not be as adept at recognizing quality toxicity studies which identify toxic endpoints. It is highly advisable that a trained toxicologist who is well-versed in risk assessment be brought into the fold to assist in identifying chemical health hazards. As mentioned in Chapter 1, proper risk assessments are carried out by a diverse team, and toxicologists should be among the team members.

2.3 DOSE-RESPONSE ASSESSMENT

Having gone through all available data sources and identified relevant hazards for a substance, the risk assessor moves on to the next step, the dose-response assessment. Whereas the hazard identification step attempts to answer the question "what can this chemical do to me?", the dose-response assessment step goes further and attempts to answer the question "to how much do I have to be exposed for the unwanted effects to occur?" It is virtually impossible to answer one question without straying into the territory of the other, so these two steps are often performed simultaneously or very closely together.

In contrast to other risk assessment efforts, such as those for environmental contamination or community exposure, the purpose of the dose-response assessment from an industrial hygiene perspective is to obtain an OEL or an occupational exposure band (OEB). As we will see, some chemicals have a wealth of information from which to draw such conclusions, while others are sorely lacking in toxicological information (data-rich substances versus data-poor substances, a dichotomy which is a common occurrence). The rest of this chapter will outline how these topics tie back into the dose-response aspect of the risk assessment and how it impacts industrial hygiene. It is important to note that this portion of the chapter is intended to be a quick introduction to the toxicological aspects of risk assessment which are of importance to the industrial hygienist.

2.3.1 DOSE-RESPONSE CURVES

Arguably the foundation for all toxicological evaluations begins with the dose-response curve. A dose-response curve is a graphical summation of all the experiments performed with a substance being administered to a test population at varying doses. Figure 2.3 shows a typical dose-response curve for a threshold substance. The graphs are set up such that the endpoint of concern (i.e., the physiological manifestation of the toxicological effect being studied) is displayed as a percent of the population in the study on the Y-axis (dependent axis), and the administered dose is plotted along the X-axis (independent axis). The dose is often plotted as the logarithm of the administered dose.

The dose-response curve tells the risk assessor a significant amount of information. One feature is the steepness of the curve; that is, the steeper the curve, the lesser the difference between a "toxic dose" and a "non-toxic dose". In other words, a substance can become more toxic over a smaller exposure compared to other substances. Such outcomes have significant impacts in industrial hygiene and occupational exposures. Another feature of the dose-response curve is that it can showcase two very important doses, the no observed adverse effect level (NOAEL) and the lowest observed adverse effect level (LOAEL). The NOAEL is the highest administered dose of a substance to a test population at which the desired toxicological endpoint did not manifest in any of the subjects. Conversely, the LOAEL is the lowest administered dose to the test population

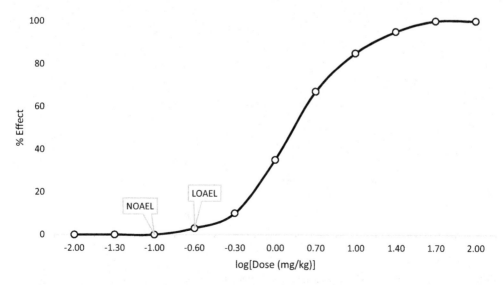

FIGURE 2.3 A typical dose-response graph.

at which the adverse effect of concern was observed in the test population. These two particular dose thresholds have classically been utilized as the primary targets for risk assessment purposes, particularly the NOAEL.

However, the use of the NOAEL as a critical safety threshold for risk assessment has been criticized.[17] Some of the more concerning features of the NOAEL are that the results are subject to the sensitivity of the test or assay being conducted. It is entirely possible that an effect occurs at a given dose but is simply not detected. In this instance, a false threshold would be assigned and skew the remaining steps of the risk assessment. Other aspects of the study design also impact the resulting NOAEL, including the doses chosen for the study, the spacing between doses, and the study duration.

An alternative to the NOAEL threshold approach that has gained considerable momentum in the toxicology field is the use of the benchmark dose (BMD).[16] This method uses a mathematical model to fit experimental data to a curve associated with a pre-determined incidence increase. The increase is usually set at 5% or 10% increase over the occurrence in a control population. In other words, if liver damage was the endpoint of interest and it occurred in the control population at a rate of 1 in 5,000, the BMD would fit a mathematical curve to the dose at which liver damage occurred in 5% or 10% increase in this background incidence rate. The fitted curve to the increased incidence rate is the 95% lower confidence interval of the curve and is referred to as the BMDL. In contrast, the 95% upper confidence interval is also fitted to the same incidence rate and is referred to as the BMDU. For risk assessment purposes, the BMDL can be utilized as the point of departure (PoD) for further calculations. A distinct advantage of this method is that it takes all experimental data into consideration, including statistical variability and uncertainty, in the formation of the fitted curve. The BMD methodology has become the preferred threshold dose technique within the EU[18] and the US EPA (Figure 2.4).[19]

We mentioned earlier that Figure 2.3 is the shape of a dose-response curve for a threshold toxicant. The term "threshold" indicates that there is a dose below which the toxicological effect will not be present. Many toxicological endpoints are considered to have thresholds, such as acute

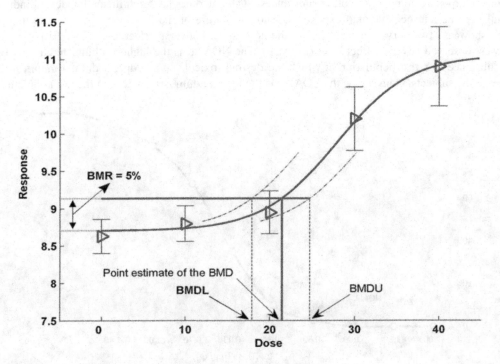

FIGURE 2.4 A benchmark dose graph. Image copyrighted EFSA.

toxicity (LD_{50}). However, some of the more concerning toxicological effects, such as cancer and mutagenicity, are considered to be non-threshold; that is, they are able to exert their effects at even extremely low concentrations. Historically, in these situations, the dose-response curve was often extrapolated to doses even lower than those tested for via experimentation.[20] Since there was no presumed threshold dose, the intention was to tabulate a dose which does not generate an excess rate risk, such as 1 in 100,000 people. A linear extrapolation was performed from the experimental point of departure (the LOAEL) through the axis origin. The slope of the linear extrapolation was then used to calculate theoretical doses at which excess relative risk was believed to be seen[21]. However, this approach has largely been supplanted by the BMD method. The utility of the BMD has been shown to be applicable to threshold and non-threshold toxicants alike.

The output of a classical dose-response curve typically arises from epidemiological studies or, more likely, from animal studies. Recalling the data types from "Hazard Identification", the former data type is typically the most difficult to come by. Consequently, animal studies are the most commonly utilized source of data for a dose-response assessment. However, looking at Figure 2.3 gives some insight to the effort required to gather the data using animal models. A total of 11 data points are presented, but each data point represents a set of experiments which utilized not only a test population of animals, but also a control population. It does not take long to see that a significant number of animals are utilized to construct a single dose-response curve. The costs to purchase, house, and care for the animals create a very significant dollar figure for the dose-response curve as well. Taken together, it is easy to see how this process does not adhere to the "3 Rs".

2.3.2 Use of In Vitro Studies

We had previously mentioned that in vitro assays and studies are becoming more prevalent for the use of hazard identification, identifying toxic endpoints much faster and more accurately compared to animal models. The entire purpose of the dose-response assessment portion of the risk assessment is to identify potentially safe doses or thresholds for human exposure. By their very nature, a dose is an amount of a substance administered to a person normalized to some unit of measure, most often body weight. In practical terms, this becomes milligrams of material administered per kilogram, or "mg/kg". The output of an animal study is that it conveniently provides data in (almost) directly applicable units.

However, as mentioned earlier, animal testing is being phased out of various industries, either by governmental regulation or by self-regulation. The obvious alternative is to utilize in vitro studies, but these assays typically identify endpoints in concentrations, such as μmol/liter (μM). It should be readily apparent that a concentration cannot be immediately or easily converted into a useable human dose. There is significant debate on how best to perform this conversion. A blossoming field is in vitro-in vivo extrapolation (IVIVE).[22] A significant amount of work is being conducted in this field for predicting pharmacokinetic and pharmacodynamic profiles for new drugs, specifically to address issues relating to mandates by the EPA and Congress. While the bulk of the effort has been toward predicting pharmacological profiles, the same process can be applied to toxicological endpoints as well.[23] These techniques provide a simulated or calculated dose based off the in vitro assay data, thereby overcoming the concentration vs. dose dilemma. If performed over a wide range of concentrations, the data can provide a calculated dose-response curve similar to what is provided from in vivo tests. However, these techniques are still relatively new and continually evolving, and many risk assessors have yet to fully embrace them as legitimate dose data. Understandably, there is a significant amount of uncertainty in IVIVE results, as there should be. As the field continues to evolve, it may very well be that IVIVE becomes a common source of data for the risk assessor.

2.3.3 Evaluating the Quality of the Study

For any given human health risk assessment, there is a distinct possibility that a multitude of studies, reviews, and articles can be identified with information that could possibly be relevant to the assessor. Database searches can retrieve numerous sources that need to be reviewed for their applicability. But not all sources provide the best data, or even useable data. For instance, a study could conceivably report a NOEAL for a particular toxicological endpoint, but if the study failed to mention the entire dosing regimen or they utilized a non-applicable model, then the utility of the finding is questionable. It is up to the assessor to review each document and study being considered for their utility. In reality, finding the studies which relay or report a hazardous toxicological endpoint is a time consuming and frustrating endeavor, but the evaluation of said studies can be even more so.

In order to address the quality evaluation gap, a method for evaluating the quality of a toxicological study was put forth by Klimisch.[24] Appropriately referred to as the "Klimisch Score", it is a systematic evaluation of any toxicological study to assess its validity. Each study is reviewed and assigned a score: 1 (reliable without restriction), 2 (reliable with restrictions), 3 (not reliable), or 4 (not assignable). Each score category has its own criteria and merits which allows for assigning of studies. The Klimisch system has found wide popularity in the pharmaceutical sector among chemical risk assessors as it allows them to systematically categorize the multitude of studies which are often found during the assessment process. An important feature of using the Klimisch scoring method is that since it is a systematic approach, it greatly reduces bias during the evaluation process and consistently rates higher quality studies with better scores, ensuring they are used effectively during the assessment process.

The implementation of a systematic categorization scheme for toxicology studies is not only useful, but it also presents the risk assessment team with the first instance of risk acceptance criteria. We mentioned in Chapter 1 that prior to beginning any risk assessment, the criteria of the assessment had to be identified and agreed upon by all parties. With four categories of study quality rating, it is incumbent on the risk assessment team to decide beforehand which Klimisch scores are useable for the risk assessment. For instance, it is not uncommon for small laboratories with fewer resources to conduct quick studies to obtain rough data points, and such studies may not strictly adhere to OECD guideline specifications. It is entirely possible that such studies would be graded lower on the Klimisch scale, but would the study be acceptable in such a case? If the risk assessment team decides to exclude studies with scores of 3 and 4, they are potentially losing access to a myriad of studies, thereby restricting the amount of data at their disposal; on the other hand, the decision to include such studies runs the risk of poorly executed experiments and perhaps irreproducible results. There is no easy answer for this particular question and is a key risk criteria question which must be addressed when performing any human health risk assessment.

2.3.4   Creating Occupational Exposure Limits (OELs)

Up to this point on the surface, the first two steps of the human health risk assessment have been completed. The hazard identification is performed and finds any relevant toxicological endpoints that may be of occupational concern. From the same studies which classify the toxicological endpoints, the risk assessor (normally the toxicologist) then evaluates the quality of the studies and evaluates the dose data to identify thresholds at which adverse effects are likely to occur. However, the dose data thus identified typically pertain to animal studies (in vivo data), are from cellular assays (in vitro data), or are modeled or predicted from various non-testing methods (QSAR, read-across, in silico). None of these formats are directly applicable to human doses nor can they be directly transferred to identify safe levels for workers. How then, are those in the pharmaceutical industry supposed to know if occupational levels are too high? The answer to this question is that the acquired dose-response data can be utilized to craft OELs.

Before delving further into the topic of OELs, it is absolutely paramount to understand what an OEL is. An OEL is an airborne concentration threshold to which it is believed that the majority of healthy, adult workers can be exposed over an 8-hour time-weighted average, 5 days a week, 40 hours per week, over the course of a working lifetime and not experience adverse health effects.

OELs are typically set by scientific and public health authorities with significant knowledge in epidemiology, public health, medicine, toxicology, and related disciplines. Contrary to common use, an OEL is not intended to be a strict delineation between "safe" and "unsafe" concentrations; it is not a "line in the sand" beyond which exposures result in guaranteed deleterious health effects. Rather, they should be used from a risk perspective in that they are thresholds below which routine exposure will not adversely affect the majority of healthy workers. The emphasis on the working population is important because an OEL is not typically crafted with sensitive populations in mind, such as pregnant women, children, elderly, or immunocompromised individuals. Finally, OELs are designed specifically for exposures in the workplace and are not considered to be protective of the general community or environmental exposures. The OELs are typically established in units of parts per million (PPM) for solvents, gases, and vapors, and in units of milligrams per cubic meter (mg/m³) for particulates, aerosols, and fumes.

A slight variation of the 8-hour OEL is a short-term exposure limit, or STEL value. It was recognized that activities in various industries would, from time to time, briefly expose employees to higher concentrations. These brief excursions are allowed to be 15 minutes in duration and must have a minimum of 1 hour between exposures, with no more than four such exposures occurring in a day. The reason for this is to minimize acute exposure to the employee while allowing enough time for the body to clear the toxicant from the system. A STEL value is often tabulated separately from the 8-hour OEL but is still factored into the full 8-hour time-weighted average.

Depending on the authoritative body which crafts the particular OEL, they can have different names. The US OSHA maintains it list as the permissible exposure levels (PELs); the American Conference of Governmental Industrial Hygienists (ACGIH) annually publishes its list of threshold limit values (TLVs); the US National Institute for Occupational Safety and Health (NIOSH) disseminates a list of recommended exposure levels (RELs). Virtually, all the major industrialized countries have some authoritative body which publishes a list of country specific OELs, and these hold sway in their country of origin. Of particular note, the publicly available GESTIS database (https://www.dguv.de/ifa/gestis/gestis-internationale-grenzwerte-fuer-chemische-substanzen-limit-values-for-chemical-agents/index-2.jsp) is a compendium of international OELs for a multitude of common industrial substances which any industrial hygienist can access to gain insight to occupational thresholds (it is not uncommon for different authoritative bodies to assign different OEL values to the same substance, reflecting a difference in interpretation of toxicological data and acceptable risk levels).

Yet with all the differing groups which evaluate hazards and dose-response data to establish acceptable airborne concentrations for the workplace, it is surprising that none have assigned such thresholds to pharmaceuticals. Common industrial substances such as metals and solvents have been used in the workplace for a multitude of decades, allowing for the accumulation of a myriad of epidemiological data and other studies of the various toxic effects of these substances. These materials will almost assuredly have OELs established. But APIs and their corresponding intermediates are materials which are intentionally designed to exert physiological effects at increasingly lower doses, rendering them far more potent than traditional industrial chemicals. It would seem prudent that these materials be given a high priority for OEL establishment, but that has not been the case with authoritative groups. Interestingly, it was the pharmaceutical companies themselves that initially identified this gap and attempted to rectify the issue via self-regulation and the creation of their own internal OELs.

A considerable amount of work has been conducted in an effort to establish scientifically based exposure levels to APIs.[25] The methods that were utilized in the late 1980s and early 1990s to calculate an OEL for an API has undergone some modifications, but for the most part has remained unchanged. The equation to calculate a pharmaceutical OEL is shown in Equation 2.1. (PoD represents a toxicological point of departure, or POD, mentioned previously; BW is body weight of a healthy human, typically considered to be 70 kg; UFc is a composite uncertainty factor; PK is pharmacokinetic adjustment factors; and V is the volume of air breathed in a typical work day, assumed to be $10\,\text{m}^3$.)

General method to tabulate API OELs.

$$\text{OEL}\ \left(\text{mg/m}^3\right) = \frac{\text{PoD} \times \text{BW}}{\text{UFc} \times \text{PK} \times \text{V}} \tag{2.1}$$

The OELs tabulated from Equation 2.1 are intended to mirror the ACGIH definition of a threshold limit value (TLV). That is, they are intended to be time-weighted average reference points for airborne concentrations to which most employees can be exposed, for 8 hours a day, and not experience adverse health effects. Such levels are not designed to be treated as strict levels of "safe" versus "unsafe"; rather, they are guidelines which allow industrial hygienists to determine exposure risks to airborne levels of the chemical.

2.3.4.1 Evaluating the PoD Criteria

Herein lies a set of criteria that must be established in the risk management process. Risk managers and stakeholders must agree upon the criteria for each variable and how it should be used. For instance, the point of departure represents a critical effect which is relevant to the working population that occurs at the lowest dose. Since an API is designed to have a biological effect, many may be tempted to simply use a therapeutic dose of a substance as the PoD. However, the intention is to protect employees from all adverse health effects, and most APIs have more than one effect. In addition, a significant number of substances do not have established therapeutic doses; that is, they are still experimental. In these cases, risk assessors must go back to the hazard ID and dose-response curves to establish an appropriate PoD.

Furthermore, there is more than one commonly used PoD in toxicological risk assessments. The traditional method has been to use the no observable (adverse) effect level, or NO(A)EL or the lowest observable (adverse) effect level, or LO(A)EL.[24] There are some substantial drawbacks to using the NO(A)EL or the LO(A)EL as a PoD in Equation 2.1. One such drawback of using the NOAEL or LOAEL as the PoDs is that they are reliant on the doses administered during the assessments. It is up to the discretion of the one performing the tests to choose the initially administered doses. In line with the current philosophy of using as few animals as possible and thus as few doses as possible, it's entirely possible (and even likely) that the administered doses are significantly above or below the true NO(A)EL or LO(A)EL of the substance. Additional experiments can be performed, but this goes against the prevailing mantra of replacement, reduction, and refinement. As we mentioned earlier, there are other drawbacks to the NOAEL approach which often provide sources of criticism. One is the study duration. Shorter studies may not provide enough time for certain toxic endpoints to manifest themselves, thus providing a false PoD. Also, the test itself may not be sensitive enough to detect the anticipated endpoint, even though it may be occurring in a statistically viable quantity. Finally, the classical NOAEL approach is not suitable for non-threshold endpoints such as genotoxicity, mutagenicity, and carcinogenesis. For these reasons, the benchmark dose method (BMD) is gaining considerably more acceptance as a suitable risk-based PoD, as described earlier in this chapter.

Despite these issues, the traditional OEL assignment for APIs still often utilizes the NOAEL from various animal studies. It is incumbent on the risk management team to seek expert advice from the resident toxicologist on the most prudent PoD to use in the OEL establishment. Not surprisingly, the choice of the PoD has a significant effect on the outcome of the OEL, so having a well-thought-out rationale for choosing one over another must be clearly laid out by the group. Whichever path forward the risk management team takes, it must be scientifically accepted as valid and also clearly documented for future assessors and auditors alike to follow.

2.3.4.2 Evaluating the BW Criteria

Depending on the source, there are several "standard" values which are used for the BW variable. Typically, the value used by assessors is 70 kg, but this can be as little as 50 kg. Since the value is located in the numerator of Equation 2.1, the larger value results in a marginally higher derived OEL, thus making it less conservative. This point of contention has been raised before from the standpoint that an acceptable "dose" for a 70 kg individual could potentially be unacceptably high for a 50 kg individual. Many risk assessors argue, therefore, that a 50 kg standard for BW should be used. However, there is no standardized method for this application, so the choice of BW criteria becomes a choice of the organization itself.

2.3.4.3 Evaluating the UFc and PK Criteria

Uncertainty abounds in designing OELs for APIs. There is a general consensus that uncertainty factors (sometimes called correction factors) need to be a critical component of the OEL calculation, but exactly how this was accomplished varied over the years. Original uncertainty factors were simply to divide a derived OEL by 100 (termed a "margin of safety"), but this was an arbitrary calculation. Over the years, many layers of uncertainty have been identified and most importantly ways to scientifically assess values to them have been developed.[26]

Today, a composite uncertainty factor (UFc) is used which is a product of all individual uncertainties surrounding a particular toxicological assessment.[27] These are summarized in Table 2.1. Each individual uncertainty factor is assessed individually based on expert judgment. Once identified, all of the uncertainty factors are multiplied together to derive the UF_C.

References which provide guidance on the exact value of the uncertainty factor to apply are provided by various regulatory bodies and organizations, including the EPA, ICH, FDA, and ECHA.[28] In-depth considerations for each uncertainty factor can be obtained from these sources, but we will briefly detail what each factor is and what it attempts to resolve. The individual susceptibility factor (UF_H) takes into account the variations between human responses. The unique genetic makeup and the spectrum of metabolic activity between persons presents a challenge in protecting nearly everyone. The UF_H is applied in an effort to protect the uppermost sensitive sector of the everyday working population. The animal to human variability factor, UF_A, is applied to account for the extrapolation of animal data to human data. As we have mentioned previously, animal data has long been recognized as not directly comparable to human responses. Allometric scaling guidance has been provided by the FDA for human equivalent dosing,[29] and similar approaches are utilized in applying safety factors to OEL establishment. The dose-response factor (UF_L) is assigned to take into account extrapolations from LOAEL to an NOAEL. It is not uncommon for studies to not arrive at a NOAEL, simply because it is a factor of the study design. In these instances, the lowest dose which is reported is, by definition, the LOAEL for that study. Additionally, if a therapeutic dose is used as a PoD, that is often treated as a LOAEL. In these instances, the UF_L is assigned to extrapolate to a dose at which the adverse effect is not observed. The exposure duration uncertainty

TABLE 2.1

Uncertainty Factors for Tabulating OELs

Factor	UF Range	Default Value
Individual susceptibility (UF_H)	1–10	10
Animal to human variability (UF_A)	1–12	Allometric scaling[41]
Dose-response (UF_L)	1–10	3
Exposure duration (UF_S)	1–10	Compound specific
Database quality (UF_D)	1–10	1

factor, UF_S, is utilized to describe the potential differences between a short-term study and a long-term study. Frequently we see studies attempting to identify endpoints which require a significant amount of time for development, yet in a human population certain endpoints may take many years to manifest themselves. The UF_S accommodates this discrepancy. The final standard uncertainty factor is to account for the quality of the data (UF_D). The use of the Klimisch score system can help feed into this factor.

In recent years, additional uncertainty factors or modifying factors have also been used in assigning OELs. These include the use of bioaccumulation factors and severity of effect factors. Bioaccumulation is not a reference to how much material builds up within the environment; rather, it attempts to describe if material accumulates within the body over time. For many industrial chemicals, such as metals, this is a very real hazard. It is not beyond the realm of possibility for an API to accumulate over time as well, especially by varying routes of exposure. A bioaccumulation uncertainty factor attempts to incorporate this feature into the OEL calculation. Similarly, a severity of effect factor has been suggested to modify an OEL based on more severe effects. However, such a modification factor has not been utilized by many within the industry when setting an OEL.

The final factor in the denominator of Equation 2.1, PK, is a modifying factor based on pharmacokinetics. In reality, there are several sub-factors which feed into this uncertainty factor, including pharmacokinetics/pharmacodynamics[30] and bioavailability.[31] Increasing evidence shows that the way a substance behaves within the body, through metabolic processes and distribution profiles, greatly impact how health hazards manifest themselves. For instance, it was originally assumed that 100% of an exposure to an employee was turned into a full-body dose. In reality, the substance undergoes significant metabolic activation/deactivation/clearance, and the body's clearance system also plays a role in what the actual dose to employees is. This becomes an extremely complicated subject which requires a vast amount of data to make an assessment, and often such data is only available only in later stages of drug development. Likewise, the original assumption was that 100% of an occupational exposure (dose) was bioavailable, but studies have again shown that this is not necessarily the case. This is further compounded by the fact that bioavailability of a substance is significantly impacted by route of administration, and since most occupational exposures occur via inhalation, bioavailability may not always be 100%. This is especially true for large, molecular weight substances like proteins or peptides, which have been shown to have significantly reduced bioavailability via the inhalation pathway.[32] In short, the use of PK factors have significantly fewer guidance points than typically utilized uncertainty factors, and consequently, these are not as frequently utilized in setting OELs, although they should be since they provide more realistic threshold levels for worker safety.

It should be readily apparent that significant professional judgment comes into play when assigning uncertainty values during the assessment. An experienced toxicologist is needed to ascertain the desired values for use in assigning the uncertainty factors. Much of the decision may be based on professional experience, or default values can be utilized throughout. Regardless, the specific uncertainty factor will vary from substance to substance, but the values utilized must be documented as to why they were chosen for their scientific basis.

2.3.4.4 Evaluating the V Criteria

The final variable in Equation 2.1 is the volume of air, V. It is standard practice to assume that an employee breathes $10\,m^3$ of air in a typical 8-hour day. The units of the volume variable, m^3/day, provide the basis of the concentration for the OEL.

2.3.4.5 Calculating the OEL

Once all the requisite information is obtained, a qualified team member then proceeds to calculate the OEL. However, it is critical to understand that all the steps mentioned up to this point are highly technical. Knowing how to appropriately identify hazards and understanding how to evaluate sophisticated toxicological studies is a very difficult assignment. Unique requirements are needed for an individual to appropriately perform the assigned task.

It should come as no surprise that the most qualified members of the risk management team to provide this input, again, should be a qualified toxicologist. The larger science-based organizations have the luxury of an in-house toxicological group that can dedicate several members to not only evaluating previously published toxicological studies, but potentially plan additional experiments if the resources are available. Such experiments can fill needed gaps and provide additional guidance in the way of establishing OELs, but such experiments cost significant time and money. In contrast, smaller organizations may not have a dedicated team or department capable of full-time toxicological analysis. In these circumstances, it is prudent to outsource such decision-making responsibilities to qualified consultants. There are numerous firms located around the world which specialize in assessing toxicological risks for APIs and related materials. Many of these firms have long years of experience working with small and mid-sized biotech firms to establish OELs of substances both new and old. Working with such consulting firms should always be viewed as a valuable addition to the risk management team.

Applying a hypothetical example can showcase how Equation 2.1 is applied. Suppose a toxicologist performed a hazard assessment and dose-response assessment for a new substance to be used in a consumer healthcare manufacturing site. The assessment identified liver damage as the most sensitive PoD with a NO(A)EL of 35 mg/kg. Applying various uncertainty factors provided an UFc of 80 (unitless), and the toxicologist noted that pharmacokinetic differences may result by route of administration and thus assigned a PK of 5 (unitless). Using the standard value of 10 m³ for V, the toxicologist calculated:

$$\text{OEL}\left(\frac{mg}{m^3}\right) = \frac{35\frac{mg}{kg} \times 70 \text{ kg}}{80 \times 5 \times 10 \text{ m}^3} = 0.613\frac{mg}{kg} = 613 \text{ } \mu g/m^3 \tag{2.2}$$

The final tabulation of the OEL is the ultimate goal and deliverable of the first two steps of the risk assessment process, and it is the goal which should be applied to every substance being handled by workers. After all, every chemical has the potential to exhibit harm at some level, but without knowing that threshold then the worker is potentially at risk. But to get to this point, a significant amount of work has been input and utilized a wealth of data and resources to arrive at this goal. In this context, we would refer to such substances as "data-rich" compounds; that is, there are numerous in vitro and/or in vivo studies to provide context to the risk assessor from which to gauge their values for the OEL. But not every compound handled has such a library of data for it. For newly identified targets, intermediates in a synthetic pathway, impurities in a batch, or perhaps non-pharmaceutical substances, there is little to no data available. Such scenarios are "data-poor" substances and provide risk assessors with significant challenges to maintaining the safety of the employees handling such materials. In light of such data gaps, how then do industrial hygienists, risk assessors, and toxicologists all work to establish safe working limits for these substances? One frequently used approach is exposure banding.

2.3.5 EXPOSURE BANDING AND OELS

Exposure banding (sometimes used synonymously with the terms control banding or hazard banding) is one which categorizes substances into distinct groups or bands. Exposure banding was first described within the pharmaceutical sector several decades ago but has evolved quite a bit in the intervening years since.[33] With exposure banding, substances are placed into bands based on where the presumed OEL would fall. It is important to note that the OEL of the substance being banded is unknown since there isn't enough data to ascertain it. However, the bands provide "best guess" as to what the threshold should be for the substance. A convenient feature of such a banding system is that it allows the risk assessors to develop a list of known, effective controls which can keep the airborne thresholds below those of the band.[34] The success of this system in the pharmaceutical industry led to several other industries adopting the practice of exposure banding and has become a useful best-practice for any industry. In the United States, NIOSH even put forth a general exposure banding paradigm which can be utilized by general industry for substances which do not have OELs.[35]

Exposure banding of data-poor substances is particularly useful from the vantage point of risk management. However, one of the first things to decide upon from a risk management perspective is the number of bands the system will contain. Many organizations utilize a 5-band system, but there is no set rule which dictates this. The number of bands is almost irrelevant, but the key component of the bands is the criteria which separates them and the required controls to meet the airborne concentrations of the band. As mentioned above, each band is separated based on allowable exposure levels. In this manner, the bands are directly reflective of the hazards of the compounds categorized within them. The bands encompass a range of allowable exposures, and so two different substances with different OELs can be classified into the same control band, thus requiring the same controls to minimize exposures even though their toxic effects or modes of actions may be completely different.

It is important to note that there is no single, universal method to perform control banding. Each organization outlines the process differently and draws different risk-based conclusions as to how to align the bands. This has given rise to dozens of different control bands being developed. Examples of such bands and their differences are shown in Figure 2.5. For the most part, many organizations have their higher exposure substances aligning fairly well; that is, those that fall into Category 1 would all have similar allowable exposure levels. However, as one proceeds farther down the line into more potent substances, the bands do not always align. For example, it is entirely possible that a substance would fall into a Safebridge Category 3 but be classified by Lonza as a Category 4 and GSK as a Category 5. The substance and the inherent hazards it possesses would not change, but the required controls for handling that substance could potentially be markedly different. This is where the organization as a whole must specifically outline the criteria for controls in each category and explicitly state the rationale for why the exposure levels are set at the way they are. This begins to define the acceptable risk levels for the organization

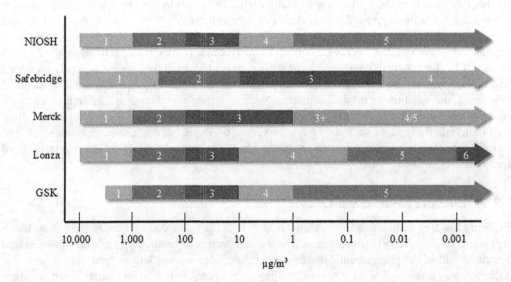

FIGURE 2.5　A comparison of selected exposure banding schemes utilized throughout the pharmaceutical sector.

while also stating in writing what the organization is committed to do to keep employee exposures to below the levels as indicated by control banding scheme.

The establishment of the bands and their respective exposure ranges may be heavily dictated from a toxicological perspective, but the resulting controls that are associated with each band will be influenced from a multitude of parties. As the name implies, exposure banding follows the established hierarchy of controls to minimize (or eliminate) exposures. For the vast majority of materials, projects, and process flows, elimination and substitution are not viable options. After all, the materials being used have been carefully designed to impart the physiological activity they do. Consequently, control banding focuses primarily on engineering controls, administrative controls, and PPE. Far and away, the preferred method of exposure control is to implement engineering controls to protect employees.

For each exposure band, significantly more stringent controls are applied to keep operators safe while handling APIs. The specific criteria for these controls must be decided upon at the conception of the risk management program and agreed upon by all parties involved. Indeed, this portion of the risk management program requires significant input from everyone. It must be communicated clearly that as the exposure bands increase, so too must the requisite controls. For those in management, they must understand that increasing engineering controls sounds fine on paper, but the capital investment can increase almost exponentially for dealing with potent compounds. For these reasons, everyone in the risk management team must come together to clearly outline what requirements they are willing to accept for each exposure band. Input from management, employees, quality, industrial hygienists, engineers, and facilities is essential so that everyone agrees on how to best tackle potential exposures. Most important, though, is that any criteria put forth at this stage must absolutely ensure that if implemented would keep employee exposures well below the designated OEL. This is one of the most difficult aspects to ensure, but a significant amount of experience in the pharmaceutical and consumer healthcare industries has shown that such levels of exposure prediction can be accomplished, and employee safety ensured. A more detailed discussion on this will be presented in Chapter 5.

It is important to note that substances are normally placed into a banding system if they have no OEL assigned to them. In this manner, there are still general guidelines for exposure control to which the practicing hygienist can test and reference. Over time, if a substance has enough test data for an OEL to be derived, then the toxicologist should work toward assigning the OEL to the compound. The OEL assignment supersedes any banding assignment from a toxicological viewpoint. Yet the utility of the exposure banding system also includes the applied hierarchy of controls for maintaining a safe working environment for the substances in the band. In this manner, it is still advantageous to the hygienist to know which exposure/control band the substance falls in so that appropriate controls can be utilized for its handling. In terms of risk management, the banding system is no longer needed for risk assessment purposes but provides valuable input for risk treatment purposes.

2.3.6 Drawbacks of Exposure Banding

The exposure banding system has been used with excellent results in the pharmaceutical and consumer healthcare industries. However, there are some notable drawbacks to its use. First, there is no universal banding system. Merck and Co. was one of the first organizations to implement an exposure banding system, but not everybody utilizes their exact banding methodology. Indeed, virtually every organization has developed their own banding system along with their own specific criteria. As a result, it can be difficult to compare banded substances between organizations. One company may establish a Category 2 material with an OEL range of 2–5 mg/m^3, and yet another group may utilize a threshold of 0.5–5 mg/m^3 for the same material. This is where justification and input from toxicological experts are critical. The establishment of such thresholds lays the foundation of the acceptable risk tolerance for an organization in regard to employee exposure. A significant amount of time and attention must be paid to establishing these limits prior to beginning the risk management process.

A further complication exists in that not every organization may use a 5-band system. The Safebridge system follows the same basic banding principles as those mentioned above but utilizes a 4-band system.[36] A common banding system utilized among contract manufacturing organizations, the Safebridge system incorporates all the same sets of criteria as systems utilizing more bands. In contrast, other organizations may utilize a 6- or 7-band system.[37] By having different numbers of bands, this can potentially alter the risk ranking of hazards and substances, which could have a significant downstream effect regarding prioritization.

Another drawback of exposure banding is one which applies to all OELs, even the TLVs. Exposure banding and the corresponding ECLs are intended to provide guidance for healthy adult populations. As mentioned before, they are levels to which it is believed these healthy populations can be exposed, as a time-weighted average, for an 8-hour workday and not experience the adverse health effects listed in that particular band. Susceptible populations such as older workers, those with underlying or pre-existing health conditions, genetic pre-dispositions, or pregnant women are not typically considered during the design of the ECLs. While an argument could be made that these subgroups of employees comprise a small minority of the overall worker population, it is the duty of the industrial hygienist to protect all employees. It is typically a significant challenge to provide protection to such workers in all industries, not just the pharmaceutical sector.

A fourth and often overlooked drawback is that the banding system categories are based solely upon employee exposure by the inhalation route of exposure. Undoubtedly, the most significant route of exposure is via inhalation, and significant efforts have been made over the years to reduce and in many cases completely eliminate the possibility of inhaling pharmaceutical powders. However, dermal exposures can potentially represent a significant route of exposure, especially to those materials which are lipophilic and can linger on surfaces for an extended period of time. There has historically been no guidance regarding acceptable amounts of materials on surfaces, but this has begun to change in recent years.[38] Various standards in other industries have brought surface contamination to the forefront of employee exposure routes, particularly the USP 800 standard in healthcare facilities in recent years.[39] In spite of these advances, most exposure banding paradigms do not take dermal exposure into account.

Still yet another drawback for the exposure banding system is relatively recent development within the industry. The banding system discussed thus far was developed for small molecule therapeutics which completely dominated the pharmaceutical sector. Within the last 10–15 years, as our knowledge of human genetics and quantum leaps in biotechnology have occurred, a significant focus on biological agents as therapeutics has emerged. These biopharmaceuticals represent a complete paradigm shift from previous materials being produced and do not neatly fall into the pre-existing banding system. A more thorough discussion on biopharmaceuticals from an industrial hygiene standpoint is presented in Chapter 3, but it is important to note that these materials are not typically categorized with traditional banding systems.[40]

A final drawback to exposure banding is perhaps the most notable and frustrating to risk assessors. Banding systems have become useful tools for assigning general exposure guidelines for substances, giving risk assessors much needed information so they can perform risk assessments on substances. After all, a human health risk assessment requires hazard identification and at least some semblance of toxicity data. However, a glaring issue becomes manifest once the exposure bands are defined; that is, how does one place a substance into a particular band? If, by definition, a substance is data poor, there isn't usually enough information to assign it to a particular band. Situations such as these are very common for brand new entities and for large compound libraries. Some approaches which were discussed can be useful for filling these gaps, though. Of particular note is the use of structural similarities for assignment purposes. In this manner, read-across methodologies are particularly useful and can be implemented for large numbers of substances.[12]

Even with these drawbacks, occupational exposure banding has represented a massive leap forward in protecting pharmaceutical employees. The establishment of categories into which substances can be placed and have corresponding airborne levels for sampling has given the industrial hygienist a considerable advantage.

2.3.7   INTEGRATION INTO THE ISO 31000 RISK MANAGEMENT SYSTEM

The above discussion hopefully laid out the logic and methodology by which the pharmaceutical and consumer healthcare industries approach the first steps of a human health risk assessment from an industrial hygiene perspective. While developed for the simple and consistent classification of active components, there is no reason why it cannot be utilized for other substances. Commonly used materials within these industries include excipients and flavors and colors. Excipients are frequently utilized as fillers and binders for pills and tablets to impart additional and favorable properties to the deliverable medicine, and as such are used in very large quantities in the pharmaceutical sector. Likewise, the consumer healthcare sector often adds flavors and colors to their products to make them more tolerable for consumption and even for marketing purposes. In this same vein, flavors and colors are typically used in large quantities. Both of these classes of materials can be evaluated and controlled using the exposure banding methodology.

Another significant advantage of the methodology is the means by which it allows for integration into the ISO 31000 risk management paradigm. As we mentioned in Chapter 1 and at the beginning of this chapter, the ISO risk assessment consists of risk identification (does a risk exist), risk analysis (what are the odds of the risk impacting the organization), and risk evaluation (what does the organization need to do about said risk). Since industrial hygiene assessments are human health assessments, and the process for these was outlined by the NRC, it is necessary for these two similar protocols to coincide and provide a useable if not identical output. By using the process outlined in this chapter, the first two steps of the NRC human health risk assessment are effectively conducted, and the first requirement of the ISO 31000 standard is fulfilled.

2.4   SUMMARY

The NRC human health risk assessment paradigm is largely adopted by the pharmaceutical industry for evaluating industrial hygiene risks to employees. The first step is hazard identification, a time-intensive endeavor involving acquisition of various studies for toxicological endpoints potentially spanning a multitude of data types. Utilizing these data, the risk assessment team must evaluate the utility and quality of the data against any pre-determined acceptability risk-based criteria set forth prior to beginning the assessment. The acceptable data are then evaluated for points of departure in an effort to derive OELs, thresholds of airborne contaminants to which it is believed the majority of healthy adult workers can be exposed as an 8-hour time-weighted average day after day for a working lifetime and not experience the adverse health effect against which it is designed to protect. This exceedingly difficult task is compounded by the fact that not all substances have sufficient studies conducted at the time of its risk evaluation, rendering it "data-poor". Exposure banding offers a unique and pragmatic solution for substances which do not have OELs, enabling risk assessors to assign these substances into practical risk categories. The bands are associated with controls which are able to keep concentrations to a manageable level. All of these tasks should be spearheaded by a trained and competent toxicologist. Implementation of these risk-based steps fulfills not only the NRC human health risk assessment paradigm but also the first step of the ISO 31000 methodology.

NOTES

1   National Research Council. (1983). *Risk Assessment in the Federal Government: Managing the Process*. National Academies Press.
2   Popov, G., Lyon, B. K., and Hollcroft, B. D. (2016). *Risk Assessment: A Practical Guide to Assessing Operational Risks*. John Wiley & Sons.

3 Torres, J. A. and Bobst, S. (Eds.). (2015). *Toxicological Risk Assessment for Beginners*. Switzerland: Springer.

4 Erhirhie, E. O.; Ihekwereme, C. P., and Ilodigwe, E. E. (2018). Advances in acute toxicity testing: strengths, weaknesses and regulatory acceptance. *Interdisciplinary Toxicology*, *11*(1), 5–12.

5 Ferdowsian, H. R., & Beck, N. (2011). Ethical and scientific considerations regarding animal testing and research. *PloS One*, *6*(9), e24059.

6 National Research Council. (2007). *Toxicity Testing in the 21st Century: A Vision and a Strategy*. National Academies Press.

7 Bal-Price, A., & Jennings, P. (Eds.). (2014). *In Vitro Toxicology Systems*. Totowa, NJ: Humana Press.

8 https://www.science.org/content/article/us-epa-eliminate-all-mammal-testing-2035#:~:text=The%20 U.S.%20Environmental%20Protection%20Agency,studies%20on%20mammals%20by%202035. Accessed September 2022.

9 Raies, A. B., & Bajic, V. B. (2016). In silico toxicology: computational methods for the prediction of chemical toxicity. *Wiley Interdisciplinary Reviews: Computational Molecular Science*, *6*(2), 147–172.

10 Rim, K. T. (2020). In silico prediction of toxicity and its applications for chemicals at work. *Toxicology and Environmental Health Sciences*, *12*, 191–202.

11 Graham, J. C., Trejo-Martin, A., Chilton, M. L., et al. (2022). An evaluation of the occupational health hazards of peptide couplers. *Chemical Research in Toxicology*, *35*(6), 1011–1022.

12 Escher, S. E. and Blitsch, A. (2021). Read-across Methodology in Toxicological Risk Assessment. *Regulatory Toxicology*, 525–538.

13 Schultz, T. W., et al. (2015). A strategy for structuring and reporting a read-across prediction of toxicity. Regulatory *Toxicology and Pharmacology*, *72*(3), 586–601.

14 *Read-Across Assessment Framework (RAAF)*, European Chemicals Agency, 2017.

15 United Nations Economic Commission for Europe Secretariat. (2011). *Globally Harmonized System of Classification and Labelling of Chemicals (GHS) (8th ed.)*. United Nations.

16 https://www.science.org/content/blog-post/un-safety-data-sheets.

17 Crump, K. S. (1984). A new method for determining allowable daily intakes. *Toxcological Sciences*, *4*(5), 854–871.

18 EFSA Scientific Committee, Hardy, A., et al. (2017). Update: use of the benchmark dose approach in risk assessment. *EFSA Journal*, *15*(1), e04658.

19 Risk Assessment Forum. (2012, June). *Benchmark Dose Technical Guidance*. (EPA/100/R-12/001). Environmental Protection Agency.

20 Hsu, C. H. and Stedeford, T. (Eds.). (2010). *Cancer Risk Assessment: Chemical Carcinogenesis, Hazard Evaluation, and Risk Quantification*. John Wiley & Sons.

21 Risk Assessment Forum. (2005, March). *Guidelines for Carcinogen Risk Assessment*. (EPA/630/P-03/001F). Environmental Protection Agency.

22 Jaroch, K., Jaroch, A., and Bojko, B. (2018). Cell cultures in drug discovery and development: The need of reliable in vitro-in vivo extrapolation for pharmacodynamics and pharmacokinetics assessment. *Journal of Pharmaceutical and Biomedical Analysis*, *147*, 297–312.

23 1) Wambaugh, J. F., et al. (2018). Evaluating in vitro-in vivo extrapolation of toxicokinetics. *Toxicological Sciences*, *163*(1), 152–169. 2) Breen, M., et al. (2021). High-throughput PBTK models for in vitro to in vivo extrapolation. *Expert Opinion on Drug Metabolism & Toxicology*, *17*(8), 903–921. 3) Algharably, E. A. H., Kreutz, R., and Gundert-Remy, U. (2019). Importance of in vitro conditions for modeling the in vivo dose in humans by in vitro-in vivo extrapolation (IVIVE). *Archives of Toxicology*, *93*, 615-621.

24 Klimisch, H. J., Andreae, M., and Tillmann, U. (1997). A systematic approach for evaluating the quality of experimental toxicological and ecotoxicological data. *Regulatory Toxicology and Pharmacology*, *25*(1), 1–5.

25 Sargent, E. V. and Kirk, G. D. (1988). Establishing airborne exposure control limits in the pharmaceutical industry. *American Industrial Hygiene Association Journal*, *49*(6), 309–313.

26 Dankovic, D. A., Naumann, B. D., Maier, A., Dourson, M. L., and Levy, L. S. (2015). The scientific basis of uncertainty factors used in setting occupational exposure limits. *Journal of Occupational and Environmental Hygiene*, 12, S55–S68.

27 Ahuja, V. and Krishnappa, M. (2022). Approaches for setting occupational exposure limits in the pharmaceutical industry. *Journal of Applied Toxicology*, *42*(1), 154–167.

28 Sussman, R. G., Naumann, B. D., Pfister, T., Sehner, C., Seaman, C., and Weidman, P. A. (2016). A harmonization effort for acceptable daily exposure derivation - considerations for application of adjustment factors. *Regulatory Toxicology and Pharmacology*, 79, S57–S66.

29 US FDA; Guidance for Industry – Estimating the Maximum Safe Starting Dose in Initial Clinical Trials for Therapeutics in Adult Healthy Volunteers. 2005.

30 Naumann, B. D., Weideman, P. A., Dixit, R., Grossman, S. J., Shen, C. F., and Sargent, E. V. (1997). Use of toxicokinetic and toxicodynamic data to reduce uncertainties when setting occupational exposure limits for pharmaceuticals. *Human and Ecological Risk Assessment: An International Journal*, *3*(4), 555–565.

31 Naumann, B. D., Weideman, P. A., Sarangapani, R., Hu, S. C., Dixit, R., and Sargent, E. V. (2009). Investigations of the use of bioavailability data to adjust occupational exposure limits for active pharmaceutical ingredients. *Toxicological Sciences*, *112*(1), 196–210.

32 Pfister, T., et al. (2014). Bioavailability of therapeutic proteins by inhalation - worker safety aspects. *Annals of Occupational Hygiene*, *58*(7), 899–911.

33 Zalk, D. M. and Nelson, D. I. (2008). History and evolution of control banding: A review. *Journal of Occupational and Environmental Hygiene*, *5*(5), 330–346.

34 Naumann, B. D., Sargent, E. V., Starkman, B. S., Fraser, W. J., Becker, G. T., and Kirk, G. D. (1996). Performance-based exposure control limits for pharmaceutical active ingredients. *American Industrial Hygiene Association Journal*, *57*(1), 33–42.

35 NIOSH [2019]. Technical report: The NIOSH occupational exposure banding process for chemical risk management. By Lentz, T.J., Seaton, M., Rane, P., Gilbert, S.J., McKernan, L.T., Whittaker, C. Cincinnati, OH: U.S. Department of Health and Human Services, Centers for Disease Control and Prevention, National Institute for Occupational Safety and Health, DHHS (NIOSH) Publication No. 2019–132.

36 Ader, A. W., Farris, J. P., and Ku, R. H. (2005). Occupational health categorization and compound handling practice systems - roots, application and future. *Chemical Health & Safety*, *12*(4), 20–26.

37 Escopharma.comn/solutions/oel-oeb. Accessed October 20, 2021.

38 Kimmel, T., Sussman, R., Ku, R., and Ader, A. (2011). Developing acceptable surface limits for occupational exposure to pharmaceutical substances. *Journal of ASTM International*, *8*(8), 1–6.

39 <800> Hazardous Drugs - Handling in Healthcare Settings. United States Pharmacopoeia Web Site. http://www.usp.org/sites/default/files/usp_pdf/EN/m7808.pdf Accessed November 11, 2022.

40 Graham, J. C., Hillegass, J., and Schulze, G. (2020). Considerations for setting occupational exposure limits for novel pharmaceutical modalities. *Regulatory Toxicology and Pharmacology*, *118*, 104813.

41 US FDA; Guidance for Industry – Estimating the Maximum Safe Starting Dose in Initial Clinical Trials for Therapeutics in Adult Healthy Volunteers. 2005.

3 Industrial Hygiene Risk Assessment

Exposure Assessment

3.1 INTRODUCTION

In Chapter 2, we took an in-depth look at how the first two steps of the industrial hygiene (human health) risk assessment process plays out within the pharmaceutical industry. Closely following the NRC paradigm, the hazard identification and dose-response steps are heavily skewed toward the toxicological sciences. The third step, the exposure assessment, is very much in the wheelhouse of the practicing industrial hygienist as it attempts to answer the question of "how much is the operator exposed to?" The exposure assessment step continues the NRC risk assessment paradigm, thereby continuing to align with the ISO 31000 risk management methodology (Figure 3.1).

The term "exposure assessment" needs a little clarification. Since most human health risk assessments are conducted for a general population being exposed to a toxicant through a variety of mechanisms and media, there has traditionally been a large amount of uncertainty and modeling when conducting exposure assessments. One of the most authoritative bodies in this arena is the US EPA, and they have provided a significant amount of guidance on conducting exposure assessments.[1]

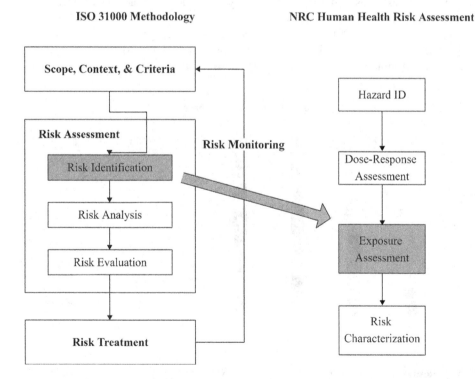

FIGURE 3.1 Exposure assessment is the third step of the NRC risk assessment paradigm and included in the risk identification step of the ISO 31000 methodology.

DOI: 10.1201/9781003273455-3

Yet such guidance is intended for assessing community exposures through all manner of exposure routes, two glaringly different aspects from occupational exposures.

It is important to note also two very different methods of assessing occupational exposures. The more accurate and time consuming is quantitative exposure assessment, which is the subject of this chapter. The other method is qualitative exposure assessment, a common tool used in various industries as a "first pass" assessment to gauge whether or not a task needs more thorough evaluation (i.e., sampling). In qualitative assessments, the practicing hygienist uses expert judgment to say based on activity, tasks performed, amount of material used, physical properties, etc. where they feel the exposure profile falls relative to the material's OEL. It is commonly utilized in many industries because this route is rapid, costs virtually nothing except the time of the hygienist, and does not disrupt operations. However, the assumption is that such qualitative evaluations are accurate. Unfortunately, a growing body of evidence is showing that qualitative assessments of exposures are routinely inaccurate.[2] Worse yet, the inaccuracy skews toward underestimating exposures; that is, employees are actually being exposed to higher levels than what qualitative assessments from practicing hygienists determine.[3] Another study found that qualitative evaluations from seasoned hygienists were often no better than the outcome from random guessing (Figure 3.2).[4] To help address these shortcomings, a number of tools have been developed to assist the practicing hygienist estimate airborne concentrations of various contaminants; however, such tools are applicable to gases and vapors. They do not necessarily apply to powders and aerosols, which constitute a significant bulk of what is handled within the pharmaceutical industry. Furthermore, given the often extremely low OELs which are assigned, it is an extraordinarily difficult task to estimate if exposures are exceeded. In these situations, quantitative exposure evaluation (sampling) is performed anyway, rendering the qualitative assessment a moot exercise. For these reasons, qualitative assessments are frequently bypassed altogether in favor of occupational exposure assessments being performed via quantitative analysis through sampling.

Occupational exposure assessment is the primary reason for the field of industrial hygiene to exist. In this manner, the practicing hygienist is concerned with exposures to a specific population with a specific health status being exposed to a specific agent (or agents) for a specific amount of time. It is a considerably different form of exposure assessment than one typically performed for environmental contamination. IH exposure assessments are highly targeted and

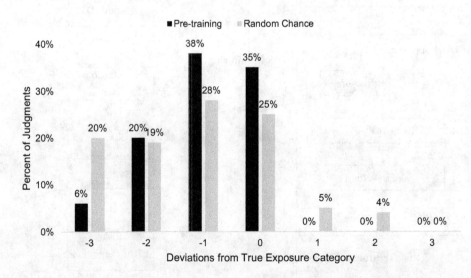

FIGURE 3.2 Results from a study on assessing personal exposures through observational means. The data show that without proper training on qualitative judgments, the results are no better than simple guessing and are heavily skewed to under-estimate exposures. Data represented from Reference 4.

specific but still require a significant amount of expertise and knowledge to be done properly. In this regard, the practicing hygienist can then formulate his or her plan on how to acquire the most useful and representative data for the risk assessment. In other words, a thorough sampling plan is required.

Many in senior management positions do not see the need for a detailed sampling plan. In truth, many hygienists in the past were told to take the bare minimum number of budgeted samples on a single day from whoever may be present so as to keep operational disruptions and costs to a minimum. This may have worked from a compliance standpoint (although in reality it almost certainly did not), but from a risk management perspective and current IH best practices, this archaic line of thinking is no longer satisfactory since it does not adequately assess employee exposures.

The entire purpose of sampling is to acquire data that is representative of employee exposures in the workplace so a proper evaluation can be performed. In order to do this effectively, a significant amount of prep work is conducted to craft a sampling plan. The sampling plan attempts to answer the following questions regarding how sampling will be performed so as to acquire the most representative data:

- **What** – for what, exactly, are we sampling? Within the pharmaceutical sector, it is automatically assumed that APIs are the target, but there is a myriad of substances which also require due consideration and should also be included in a proper sampling plan;
- **How** – what methods will we utilize for our analyses? Is the sensitivity of the method appropriate for the task duration and magnitude of the OEL? This aspect often requires a significant amount of input from the IH lab performing the analyses;
- **Where** – more precisely, where in the production process will sampling be performed, and on which tasks? As we will see, the locations and tasks for sampling on a material can be varied and more numerous than some might initially consider;
- **Who and When** – in terms the AIHA would utilize, which similar exposure groups (SEGs) are we going to sample upon? Just because two groups handle the same substance doesn't mean their potential for exposures is the same. Furthermore, folks within the same SEG can work on different shifts, which require consideration for the sampling plan as well;
- **How Many** – one of the oldest and most debated questions between hygienists and management, acquiring the appropriate number of samples is vital to give the most representative approximation of exposures while minimizing uncertainty;
- **How Much** – a well-constructed sampling plan can give insight to approximate costs not just in terms of how much the analyses will cost, but also the cost of media and how many hours will be required to execute the sampling plan.

For industrial hygienists, each one of these factors represents a significant challenge in every industry, but they can be particularly unique for the pharmaceutical and consumer healthcare sector. This chapter attempts to highlight these challenges, how the industry attempts to overcome them, and where improvement is still needed.

3.2 IDENTIFYING WHAT NEEDS TO BE SAMPLED

3.2.1 APIs AND ACTIVE INGREDIENTS

APIs and other active ingredients are the meat and potatoes of industrial hygiene sampling efforts within the pharmaceutical and consumer healthcare industries. These are the materials which have explicitly intended biological effects and from which employees must be protected against. Far and away, they represent the most important materials in any sampling plan. From a risk management perspective, APIs pose the highest degree of hazard to workers of all the materials they use and therefore deserve the most attention.

The breadth of chemical diversity among APIs is, quite simply, astonishing. These materials can range from relatively simple structures, such as ibuprofen, to complex, such as Taxol. In between these two extremes lies a plethora of chemical substances, all designed to exhibit a specific and potent pharmacological effect. These effects, while desirable for patients and consumers, are not desirable to employees handling these substances in their work environment. In addition to the known pharmacological effects, virtually every API has off-target effects (side effects) and at low doses may have significantly deleterious effects which manifest in workers. Steroids, all derived from the cholesterol skeleton, display hormonal activity and as such are targets for the endocrine system. They also play a role in inflammatory pathways, and as such exhibit activity toward the immune system. Examples of steroids include hydrocortisone, estradiol, progesterone, and fluticasone, to name a few. Antibiotics consist of varying classes themselves, such as fluoroquinolones (e.g., ciprofloxacin), beta-lactams (e.g., penicillin G), and tetracyclines (e.g., doxycycline) are all well-known substances capable of warding off bacterial infections. Yet they also possess potent sensitization capabilities, leading to severe allergic reactions among some employees. The number of chemical classes and categories are indeed numerous, and with over 100 years to study and evolve chemical structures, the sheer number of chemical structures is truly bewildering.

For hygienists working in primary pharmaceutical sites, knowledge of the API is only one issue. At such sites, if the API is manufactured, the hygienist must be concerned with chemical intermediates and potential byproducts. In a synthetic sequence, at what point does the molecule begin to exert biological activity? This is a key question which requires an answer prior to initiating a synthetic sequence.

Regarding assigned OELs, most people (including non-scientific managers) assume that pharmaceutical APIs are significantly more potent than substances found in OTC products. After all, materials which require a prescription must be more effective than those found in consumer healthcare products which are deemed "generally regarded as safe and effective" (GRASE). This is far from the truth, however. There are many substances which are available in both prescription and OTC form. Figure 3.3 shows several substances which are available both as prescription strength and as a consumer healthcare product. The significant difference is the dose being administered which dictates the need of a prescription or not, but this has no bearing on any risks imposed to the worker who makes the product.

The fact that many active substances are used in both industries raises a critical point, which is that no matter if the substance is being intended for a prescription product or an OTC product, the toxicological effects of the pure substance remain the same and workers in both industries are exposed to identical hazards. It is therefore incumbent on the proper implementation of the risk management protocol to adequately assess the controls of these materials. The hazards should be the same in both industries, and therefore, the assigned OELs should be similar (not necessarily identical, but within the same order of magnitude at least). This single fact should be enough to validate the need for routine industrial hygiene surveillance in both industries, but in the consumer healthcare industry in particular.

Knowing which API or active material for which to sample is only the first hurdle. One of the primary challenges to sampling for APIs is finding properly accredited labs to analyze said materials. As with all industrial hygiene analyses, hygienists look for labs which have been properly accredited through the AIHA-LAP (laboratory accreditation program) and whose capabilities include pharmaceuticals. There are several AIHA-LAP-accredited labs, but very few have the capability to analyze pharmaceuticals. At the time of this writing, there were only nine labs accredited to perform work with APIs, and three of those labs were housed within pharmaceutical companies (US Merck [MSD], Abbvie, and Bristol-Myers Squibb have their own accredited laboratories). Consequently, there are only a handful of groups which even have the capability to analyze these pharmacologically active substances. It is vital to establish a working relationship with these labs because they have the ability to create specific methods for APIs.

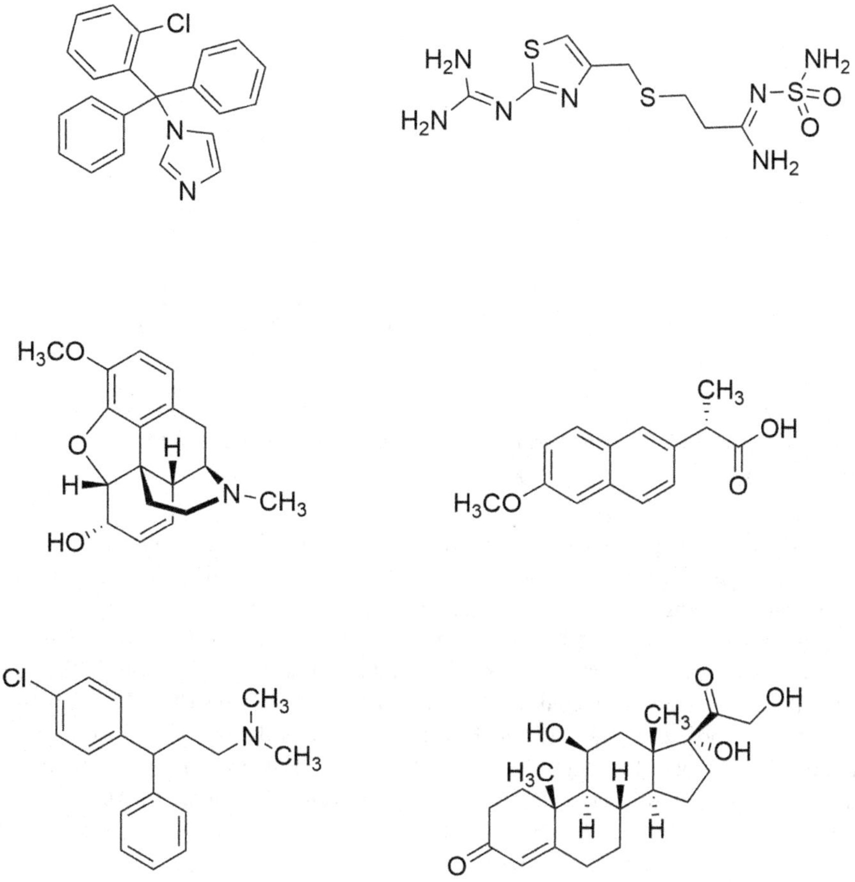

FIGURE 3.3 A collection of APIs which are available as both prescription strength and over-the-counter. Clockwise from top left: clotrimazole, famotidine, naproxen, hydrocortisone, chlorpheniramine, and codeine.

3.2.2 EXCIPIENTS

Excipients are substances found in pharmaceutical and consumer healthcare products which are not intended to impart any physiological effect within an administered drug. In essence, excipients are everything found in a product except for the active ingredient. These materials are a vital component to healthcare medicines and serve several purposes including:

- Assist in the drug delivery system by enhancing bioavailability or extending the release of the drug in the bloodstream;
- Assist in extending the shelf-life of the drug during storage;
- Prevent caking during manufacture;
- Impart additional physical properties such as hardness or softness to the final product;
- Enhance flavor or color of the final product (see the 'Flavors and Colors' section).

From a manufacturing perspective, selecting the right excipient is crucial to imparting the right properties in the final product. As such, there are hundreds of excipients in use, each serving its own unique purpose and role.[5] Some of the more common excipients include dextrose, sucrose, talc, magnesium stearate, mannitol, cyclodextrin, cellulose (and cellulose derivatives), povidone, starch, stearic acid, adipic acid, water, and gelatin. This list is by no means exhaustive (Figure 3.4).

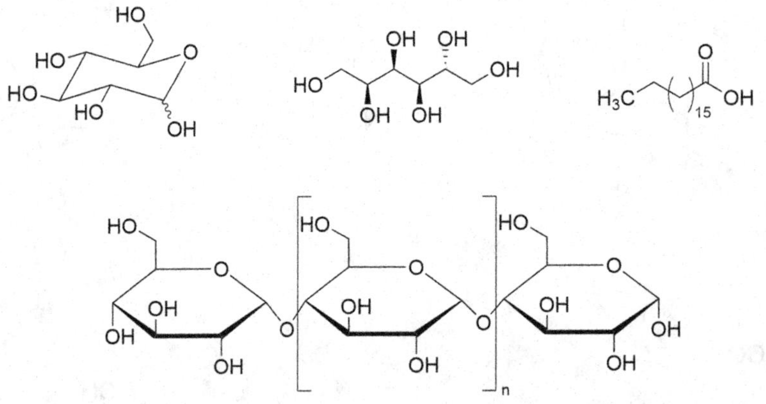

FIGURE 3.4 A sampling of common pharmaceutical excipients. Clockwise from top left: dextrose, sorbitol, stearic acid, and starch.

Excipients often make up the bulk of a pharmaceutical formulation, and by extension are used in very large quantities in manufacturing environments. Furthermore, they are often added at many points in the manufacturing process, providing multiple avenues of potential exposure to employees. From an industrial hygiene and risk management perspective, excipients have been largely ignored in comparison to APIs because they are "inert" or "not pharmacologically active". While this may be true at the doses being administered in the final product, workers handling large quantities have the potential to be exposed to amounts which would far exceed the recommended daily dose. For this reason, it is important to thoroughly treat each excipient in the manufacturing process with the same risk assessment process that APIs must undergo. Given that a single pharmaceutical or consumer healthcare product can have many times the number of excipients as the active product, this can represent a significant amount of work.

One important trend to note is that excipients tend to have higher OELs assigned to them. Whereas APIs will often have an OEL of 100 $\mu g/m^3$ or less, excipients can have an airborne limit in excess of 2,000 $\mu g/m^3$ or higher. These higher OELs often fall into the first tier of a control banding scheme, requiring less stringent controls for powder and dust exposure. Such high levels are often attributed to the fact excipients are inert, which is a valid line of logic, but documentation should still be obtained to back up any claims of an OEL. The primary issue is that very little toxicological information is available regarding even the most common excipients. Furthermore, they have been used for so long in commercial processes that it is difficult to convince researchers to conduct studies to acquire data. For new excipients being introduced in a product, it must be demonstrated that it is safe at the dose being administered in the final product, not necessarily in the handling of the material by the worker. Excipients should always be included in any sampling plan and efforts made to minimize exposure within reason.

A significant hurdle toward that goal, however, is that there are very few validated industrial hygiene sampling methods for excipients (see the section "Identifying How Sampling Will Be Performed: Method Selection and Analytical Sensitivity"). For traditional industrial hygiene applications, there are analytical methods for specific substances, but as we mentioned earlier, there has been little interest and/or need to have methods developed for excipients. This attitude is slowly reversing, since most excipients possess other hazards (such as combustibility from a Process Safety viewpoint) which need to be controlled. Regardless, the lack of method availability presents a truly significant issue for the practicing hygienist within these industries. A common work-around is the use of non-specific methods such as gravimetric analysis, but as we will see later these solutions have their own substantial shortcomings.

3.2.3 FLAVORS AND COLORS

Flavors and colors are technically lumped into the category of excipients, but they are so universally utilized (especially in the consumer healthcare sector) that they should get their own designation. Flavors and colors are used to enhance the appearance of a final product, to provide a reasonable hue for a flavored product (cherry-flavored products typically appear red, lemon-flavored materials are often yellow, etc.), and to provide flavor and adjust the level of sweetness to a product.[6] Additionally, colors are often added to final dosage forms of API to aid in differentiation among the numerous available drugs, allowing for easier identification.[7] Figure 3.5 shows a sampling of FDA-approved colors, and Figure 3.6 shows a selected number of FDA-approved flavors.

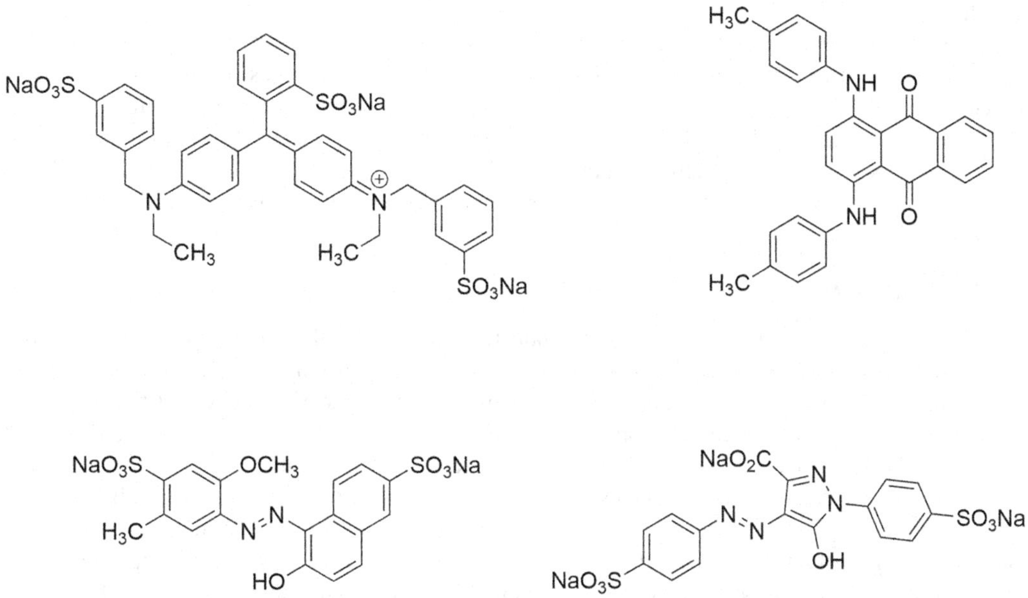

FIGURE 3.5 Examples of commonly used FDA-approved colors. Clockwise from top left: brilliant blue FCF, quinazirine green SS, tartrazine, and allura red AC.

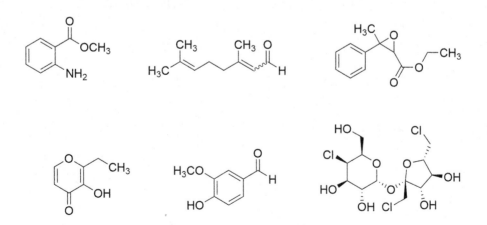

FIGURE 3.6 Examples of commonly used FDA-approved flavors. Clockwise from top left: methyl anthranilate, citral, ethyl methylphenylglycidate, sucralose, vanilin, and ethyl maltol.

Just as other excipients typically do not have specific industrial hygiene methods developed for their analysis, flavors and colors have the same problem. Non-specific methods are most often utilized to assess for airborne levels of these materials. Consequently, the same drawbacks arise in that efforts to appropriately characterize airborne concentrations of these substances are susceptible to interference. Such limitations significantly hinder the risk management process for accurately assessing exposures. Given how frequently flavors and colors are used, it is surprising that specific methods have not been developed. A distinct possibility for this is that flavors and colors have historically been classified as "safe", and therefore, they have not been considered substances of concern for operators. A recurring theme is that what is safe in a final product dose is not the same as that experienced by workers. Just as other excipients should not be ignored or discounted, neither should flavors and colors, and these substances should be present in any sampling plan.

There are additional challenges regarding flavors and colors for the practicing hygienist. In recent years, flavor and color suppliers have begun selling generalized "flavor types" rather than individual ingredients. These are often provided as "citrus flavor", "berry flavor", and others. A review of the ingredients does not provide any information as they are often proprietary. Without knowing what the components are and how much they constitute the powder mixture, it is exceptionally difficult to ascertain exposure levels using anything other than non-specific methods.

3.2.4 BIOLOGICALS

For most of the history of the pharmaceutical industry, small molecules have reigned supreme. Yet over the last 20 years, biological pharmaceuticals (biopharmaceuticals) have taken the industry by storm. Figure 3.7 shows a distinct trend in this regard.[8] It is worthwhile to note that there is not a single, industry-wide accepted definition of what constitutes a biopharmaceutical, but the FDA and Center for Biological Evaluation and Research (CBER) have issued guidance as to what constitutes a biopharmaceutical which includes blood, blood products, vaccines, somatic cells, tissues, allergenics, recombinant proteins, sugars, and even nucleic acid derivatives.[9] While small molecule pharmaceuticals still represent the bulk of FDA approvals, since 2013 at least 1 in 5 approvals has been a biopharmaceutical. Many expect this trend to not just continue, but at some point have biopharmaceuticals overtake small molecules in terms of overall approvals.

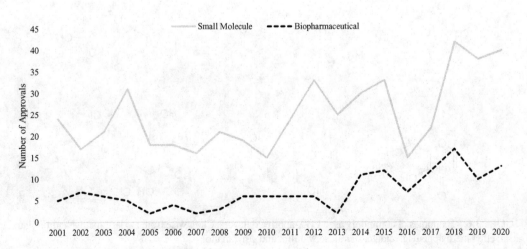

FIGURE 3.7 Trend in annual FDA new drug approvals.

There are many potential reasons for this trend. Biopharmaceuticals offer extremely targeted therapies with reduced side effects. From the patient perspective, these two properties alone make biopharmaceuticals extremely desirable and consequently patient demand for them has increased over the years. Additionally, many of the therapies have dramatically extended biological half-lives in patients when compared to small molecules. Traditional small molecule therapies typically undergo metabolism quickly and are excreted from the patient, requiring frequent dosing. Biopharmaceuticals, however, are not as readily cleared from the patient's system; in contrast, a single administration of a biological drug can last for a week or more, reducing the number of doses a patient must endure. When taken together, these offer significant advantages which the patient (i.e., the consumer) wants (Figure 3.8).

While these reasons have undoubtedly contributed to the industry-wide attention to the field, the monetary return of the medicines is perhaps a bigger driver. In 2020, 50% of the top ten revenue-generating pharmaceutical products were biopharmaceuticals.[10] The top selling drug in that year was Humira®, a biopharmaceutical which became the first API to generate over $20 billion USD in annual revenue. The return on investment for a biopharmaceutical is potentially significant, and as such, a considerable amount of R&D expenditure has been invested to develop new lines of these drugs.[11] At the time of this writing, hundreds of clinical trials were either actively ongoing or recruiting for biopharmaceuticals, including monoclonal antibodies and vaccines, indicating a substantial investment from established pharmaceutical companies.

The rise of biopharmaceuticals represents a paradigm shift for the practicing industrial hygienist. The majority of biopharmaceuticals are proteinaceous in composition and thus have significantly different properties than traditional APIs. These can include proteins/peptides and oligonucleotides, but of primary importance are monoclonal antibodies (MABs). In contrast to small molecule APIs, biopharmaceuticals have significantly different behaviors. In terms of sheer size, their mass is exponentially larger than small molecules, giving rise to vastly different properties such as solubility and stability. Their highly polar nature also makes them unsuitable to most analytical methods to which the practicing hygienist may be accustomed such as GC or HPLC. Often biopharmaceuticals require very

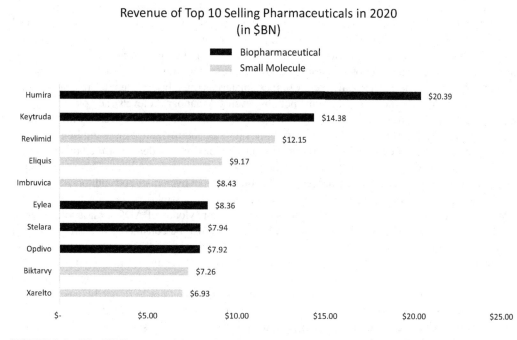

FIGURE 3.8 The 2020 revenue of the top 20 selling pharmaceuticals broken down between small molecules and biopharmaceuticals. Source: https://www.fiercepharma.com/special-report/top-20-drugs-by-2020-sales.

different analytical methods, such as the enzyme-linked immunoassay sandwich assay (ELISA). These specialized analytical methods are not utilized in all IH labs, thus making it a particularly difficult task to identify labs which craft methods for the substances. Still yet another dilemma is how one determines an occupational exposure limit for such materials.[12] Since they are essentially targeted therapies but have such different properties, do traditional OEL approaches apply? A unique banding system for protein pharmaceuticals has been proposed to address this topic but has not been formally adopted by all organizations. Despite these difficulties, biopharmaceuticals are a class of APIs with which hygienists within the pharmaceutical sector must become familiar. It is entirely plausible that this class will come to dominate the pharmaceutical sector, and it will become common for sites to be retrofitted to produce them alongside traditional pharmaceutical agents.

3.2.5   ANTIBODY-DRUG CONJUGATES (ADCS)

If highly potent small molecule APIs (HPAPIs) and biopharmaceuticals are standalone entities in the pharmaceutical universe, where they overlap would be the unique and exciting field of antibody-drug conjugates.[13] These substances are notable in that they increase the therapeutic index (sometimes referred to as the margin of safety) of APIs. Traditional medicines have a narrow range between a minimum efficacious dose and a maximum tolerated dose (MTD). Typically, the more potent a substance becomes, the narrower the therapeutic index becomes as well. ADCs circumvent this issue by delivering potent therapeutic agents to specific cells (often cancer cells) through targeted receptors, thereby minimizing (or alleviating) adverse systemic side effects. It is an exciting field of therapeutic treatment, and one which has seen an increase in activity in recent years.

ADCs consist of three distinct regions: an antibody, a linker, and a highly potent payload, or "warhead". The antibody is the source of the specificity. It is designed to bind to specific receptors which are overexpressed on cancer cells, making them ideal targets. From the perspective of the industrial hygienist, the same hazards apply here as for other antibodies. The linker is a chemical chain which, as the name implies, connects the antibody to the potent compound payload. An entire field of research has been devoted to fine-tuning ADC linker technology.[14] The linker must have significant stability to allow the entire ADC to survive long enough to reach the desired cellular destination, but labile enough to release the payload when required. It is a fine balance and exquisite execution of chemical know-how. Finally, the payload is the "active" component of the ADC, offering the therapeutic effect. The payloads utilized deserve additional attention, and deservedly so. The compounds utilized as payloads are some of the most potent cytotoxic substances ever tested. Examples of some of these are shown in Figure 3.9. The exceptional potency of these materials often brings them to single nanogram levels of allowable airborne concentrations, possibly even sub-nanogram levels (i.e., picograms/m³). Given the extreme hazards associated with these materials and their resulting extremely low OELs, it is surprising to find that very few ADC payloads have validated industrial hygiene sampling methods. This enigmatic reality of extremely hazardous materials coupled with a dearth of methods to ensure direct operator safety is a massive hurdle for hygienists working in such facilities. This issue is often overcome by the use of surrogate powder testing of equipment, but hygienists and auditors often prefer to assess worker exposures using "live" substance data whenever possible.

While particular attention has been justifiably paid to the payloads of ADCs,[15] it is vital to note that the other components, the antibody and the linker, must also be considered in a sampling plan to evaluate employee exposures. Furthermore, it is important to note that at various stages of ADC manufacturing, the toxicological properties of the materials could change. Linkers must be evaluated for their effects on individuals as well as the payload-linker unit. The final conjugation product to the antibody, even though it has highly potent payloads on it, may not necessarily have the same OEL as those payloads. This could be affected by the drug-to-antibody ratio (DAR) and is a critical component of evaluating not only the efficacy of the product but also its safety. All of these factors must be considered when constructing the sampling plan within an ADC facility and knowing at which stage of the manufacturing process the samples are taken.

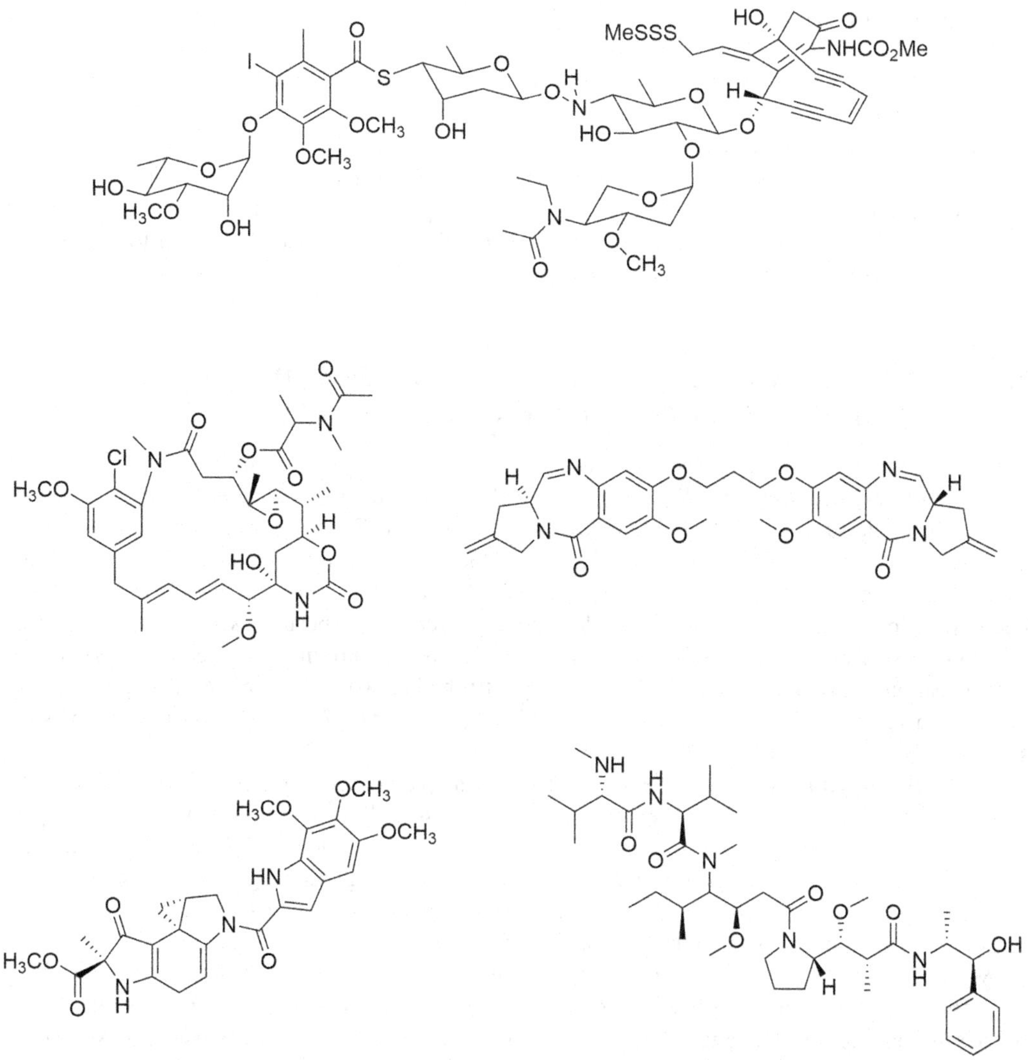

FIGURE 3.9 Chemical structures of selected ADC payloads. Clockwise from the top: calicheamicin, SJG-136 (pyrrolobenzodiazepines), monomethyl auristatin E (MMAE), duocarmycin A, and maytansine.

3.2.6 OTHER SUBSTANCES OF CONCERN

Pharmaceutical-related materials are not the only substances with which an IH must concern himself or herself. Just as traditional industries are concerned with solvents, metals, and vapors, so too is the IH in the pharmaceutical and consumer healthcare industries. All aspects of normal manufacturing operations need to be included in an overall sampling plan. This is especially true at primary manufacturing sites where APIs are being synthesized from simpler building blocks or extracted from natural sources. Such operations require copious amounts of solvents and reagents possessing a myriad of potential health effects. The practicing industrial hygienist is often kept quite busy in such settings.

Knowing the operations of the site in which the hygienist works pays dividends in this regard. For instance, for every manufacturing operation, there is most likely a cleaning operation or maintenance task associated with it. These tasks may involve the use of cleaning agents, detergents, or lubricants (not to mention all of the other substances already mentioned, including APIs).

All chemicals used at the site must undergo the same scrutiny and risk evaluation as those which have more potent effects. Being able to identify these processes and tasks can be daunting, but reading through SOPs and discussing operations with personnel is key to finding potential risk gaps which need to be assessed.

From a larger view, there is likely to be more non-pharmaceutical substances than APIs or excipients. While this shouldn't come as a surprise, it is often the case that non-pharmaceutical substances are paid significantly less attention and are given a lesser priority. It is incumbent on the practicing hygienist to advocate for these processes to be thoroughly evaluated and assessed to completely cover all employees under the risk management system.

3.3 IDENTIFYING HOW SAMPLING WILL BE PERFORMED: METHOD SELECTION AND ANALYTICAL SENSITIVITY

From an industrial hygiene perspective, we have historically been concerned with an employee's 8-hour time weighted average (TWA) exposure to chemicals; that is, what does their average exposure to a specific substance look like over an entire work shift? This approach is more in line with a compliance frameset and is often treated as such. The simple solution is to utilize a single sample for the entire shift, and the results provided are used to assess whether or not the employee has been over exposed. But this methodology does not consider peak exposures which place employees at the greatest risk. In other words, an employee could have an extremely high exposure for a very short duration in excess of agreed upon levels and still be below the 8-hour TWA. Aligning an IH program to a risk management methodology requires analysis of each individual task wherein exposures can potentially occur.

Ideally the hygienist would be able to craft a sampling plan that would allow for analysis of both task-based exposures and full-shift exposures. The most straightforward means to do so is to perform integrated sampling. Acquiring a series of samples on an employee in a single shift and then being able to average the results over the amount of time sampled gives the greatest potential insight for peak exposures. But performing task-based sampling presents another significant challenge for hygienists: overcoming the analytical sensitivity of the chosen analytical method.

As mentioned in Chapter 2, occupational exposure levels for new APIs continue to fall and are often less than 10.0 µg/m³. It is not uncommon for potent compounds to have assigned OELs in the nanogram per cubic meter range, as low as 1.0 ng/m³ in some extreme cases (such as ADC payloads). The sensitivity of the method then becomes a critical factor in being able to identify with any degree of certainty if employees have been exposed to these materials. The need for high detection capabilities (read: low quantification limits) is dictated by the amount of air that needs to be drawn through a sampler to acquire a quantifiable signal on the detector and goes hand in hand with substances having low OELs.

This relationship can be showcased with an example. Assume that a pharmaceutical site is working with a substance with an internally derived 8-hour OEL of 5 µg/m³ and short-term exposure limit (STEL) of 12 µg/m³. The plant hygienist is tasked with evaluating the risk to a worker during a weighing operation which takes approximately 15 minutes to complete. The analytical method that will be utilized by the laboratory gives a sensitivity of 1.0 µg. To determine the minimum amount of air to sample, the following equation is used:

$$\text{Air Volume (L)} = \left(\frac{A_S \ (\mu g)}{\text{OEL} \ (\mu g/m^3)} \right) \times \frac{1,000 \ \text{L}}{1 \ \text{m}^3} \tag{3.1}$$

In Equation 3.1, A_S is analytical sensitivity, sometimes referred to as the reporting limit. This represents the minimum amount of material needed to achieve a statistically viable reading on the analytical instrument (assuming complete desorption from the filter). The unit for A_S is often given as micrograms (μg). Following our example, Equation 3.1 provides a needed air volume of:

$$\text{Air Volume (L)} = \left(\frac{1.0 \ \mu g}{12.0 \ \mu g/m^3} \right) \times \frac{1{,}000 \ \text{L}}{1 \ m^3} = 83 \ \text{L} \tag{3.2}$$

In practical terms, this means that at the expected airborne concentration of 12.0 μg/m³, a minimum of 83 L of air must be sampled to collect enough material on the filter to be reliably quantitated by the analytical method. For our example, if we assume the task only requires 15 minutes to complete and the sampling method requires a pump flow rate of 2.0 L/min⁻¹, then only 30 L of air will be sampled. In this particular scenario, the filter would not have enough material on it to get a reliable signal in the analysis, and the result would likely come back as reading below the limit of quantitation. Such results are considered to be "censored data" and have historically been difficult to deal with for a couple of reasons. First, it's not possible to run traditional statistics on censored data, and second, management personnel often interpret such results as "there's nothing there" and assume the result is indicative of a safe environment.

While censored data can now be easily handled through Bayesian analysis (see Chapter 4), it's still preferable to have non-censored results. There are two ways to circumvent this shortcoming. The first is obvious: sample for a longer period. This would allow the hygienist to sample enough air and acquire enough material on the filter to get a reliable result. However, the drawback should be equally obvious. From a risk management perspective, we are interested in addressing the risk specific to the task at hand. To sample for a longer period would potentially skew the results, especially if once the weighing task is done the operator performs a completely different task with different exposure potentials. When handling potent compounds, it's vital to understand the risk associated with each task in a process to fully protect the employees. This is one of the primary tenets of risk management.

The alternative to overcoming the limitation is to develop a more sensitive method for the material being handled. For the practicing hygienist this is far and away the more desirable option and is why knowing the sensitivity of the method is so crucial prior to sampling. The sensitivity of the instruments must be enough to be able to detect increasingly lower levels of analyte in an assay. Thankfully instrumentation has advanced enough in recent years that analyses can often be performed on nanogram and sub-nanogram quantities. Another variable which contributes to the efficacy of the analysis is the material being evaluated itself. Most APIs are not volatile; that is, they do not vaporize and enter the gaseous state very easily. Consequently, APIs typically cannot be evaluated by gas chromatography (GC). Liquid chromatography (LC) and high-performance liquid chromatography (HPLC) methods are the norm for these materials along with a variety of detection methods such as UV-vis, fluorescence, and mass spectrometry.

3.3.1 Surrogates

The cost to develop a highly sensitive and specific analytical method can be significant. Many organizations may not be able to afford the required price tag to develop a method, especially if the company is a smaller biotech firm or a contract manufacturing organization (CMO) that is only being contracted for a short time to produce or package a material. In the absence of a sensitive method and the means to produce one, using surrogates presents a viable alternative for the hygienist.

Surrogates are materials for which highly sensitive industrial hygiene analytical methods have already been developed which can be substituted in place of active compounds. Surrogates are commonly used throughout the pharma world and are most often used for equipment validation (see Chapter 7).[16] The ideal surrogate should possess several key features including virtual non-toxicity (inert or innocuous substance), appropriate particle size, appropriate dustiness relative to the parent material, inexpensive, readily available, and a highly sensitive method already developed. The most common surrogates used are:

- Lactose
- Mannitol
- Naproxen sodium
- Sucrose
- Riboflavin
- Acetaminophen
- Insulin

The reporting limit for these surrogates is in the nanogram range and can even be as low as the picogram range. This offers at least 10^3 or greater enhancement in sensitivity, making them ideal for testing operations with very low OELs. Choosing the appropriate surrogate is crucial to getting relevant and useful data. It is important to know the properties of the parent substance and choose a surrogate that best mimics them. For instance, if the parent material is typically quite granular with a typical particle size of 150 µm, it would not be prudent to choose a finely divided or micronized surrogate as this would be dustier and more difficult to handle, thereby leading to inaccurate data. Lactose, mannitol, and sucrose have a distinct advantage in that they are commonly used excipients and may already be present in a mixture to be sampled. The remaining options in the list are APIs in their own right, and their use would most likely require a validated cleaning protocol to ensure cross-contamination is not occurring (always consult with quality prior to using surrogates in a cGMP setting).

Even with obstacles such as choosing a viable material and difficulty implementing a new material into a production line, the use of surrogates is very attractive. If lactose could be used as a viable surrogate for the previous example, the advantages to the hygienist would be significant. Lactose has a reporting limit of 2.5 nanograms (ng), which is equivalent to 0.0025 µg. Substituting into Equation 3.1, the required air volume becomes:

$$\text{Air Volume (L)} = \left(\frac{0.0025 \ \mu g}{12.0 \ \mu g/m^3} \right) \times \frac{1,000 \ L}{1 \ m^3} = 0.21 \ L$$

(3.3)

The new required air volume of 0.21 L is a massive improvement from the previously required 83 L and offers the hygienist a means to assess the exposures of the operation more accurately. In lay terms, if lactose were present at 12.0 µg/m³ in the air, then the hygienist would only need to collect 0.21 L, or sample for less than a minute at a flow rate of 2.0 L/min⁻¹ in order to collect enough material for analysis. Another way to view this is that for the entire sampling period, the hygienist can be assured that he or she would not get censored data from their sampling efforts.

3.3.2 Non-Specific Methods

While potent APIs require analyses with very sensitive methods, especially those categorized in higher OEB tiers, not every material receives such attention. We previously mentioned that substances such as excipients and flavors and colors more often than not do not have specific industrial hygiene sampling methods developed and thus alternatives must be sought.

One way around this dilemma is through classical gravimetric analysis of total inhalable dust exposure. A simple technique, this protocol requires the analytical laboratory to weigh a filter prior to its use and then again after a sampling protocol has been executed by the hygienist. The mass difference between the two weighings gives the total amount of powder to which the employee was exposed. This is a common published technique, such as those found in NIOSH 0500 or MDHS 14/4 and can be implemented for any substance for which a specific method is not available. However, there is a significant drawback. Since the method is non-specific, any dust or powder that is caught on the filter will be counted toward the total mass, even if it's not the material of interest. This potential interference cannot be excluded from the weighing process, and thus presents a significant point of concern for the practicing hygienist. For this reason, gravimetric analysis is typically only used for assessing "nuisance dust", or general environmental dust. It is usually not advisable to employ gravimetric analysis for multiple substances or mixtures, although it does still happen.

A second and perhaps more important point of concern is the sensitivity of the method. Depending on the analytical capabilities of the laboratory, the reporting level of gravimetric analysis can range from 20 up to 100 μg. As was shown previously, methods with low sensitivity tend to give significantly higher exposure results for task-based sampling. Suppose there is a blending task in which operators are required to charge an excipient into a reactor vessel. The task takes 15 minutes, and the sampling method (MDHS 14/4, in this case) requires a sampling rate of 2.0 L/min^{-1}. Furthermore, the company utilizes an assigned STEL of 2,000 μg/m^3, and the reporting laboratory has a method sensitivity of 100 μg. Utilizing Equation 3.1, the required sampling volume would be:

$$\text{Air Volume (L)}=\left(\frac{100\ \mu g}{2{,}000\ \mu g/m^3}\right)\times \frac{1{,}000\ L}{1\ m^3}=50\ L \tag{3.4}$$

Yet again, we have a required air volume that will not be met based on the duration of the task. For our example, the hygienist would end up sampling approximately 30 L of air, and if not enough dust is acquired on the filter, the laboratory has to report based on the reporting limit, in this case 100 μg. The final reported concentration would be:

$$\text{Concentration}\left(\mu g\middle/m^3\right)=\frac{100\ \mu g}{0.03\ m^3}=<3{,}333\ \mu g\middle/m^3 \tag{3.5}$$

The reported value is "< 3,333 μg/m^3" because there wasn't enough material on the filter to be statistically significant based on the method. As with the previous example, if there isn't enough material to be analytically sound, then the result will always be less than detectable. In this case, the end result is one which does not afford the hygienist a very usable data point for assessing the risk of the task to the worker. If the controls around the operation were working as intended and the actual airborne concentration of the excipient was 120 μg/m^3 (well below the task-based OEL of 2,000 μg/m^3), the hygienist would not be able to adequately assess the controls based on the duration of the task. This is a common problem that practicing hygienists encounter in the pharmaceutical sector and can be applied to almost every excipient already mentioned in this chapter.

Such issues are common for task-based sampling of very short duration. Indeed, Figure 3.10 shows the effect that sampling time has on the outcome of the data for non-specific gravimetric analysis. For the chart, the data were calculated using a flow rate of 2.0 L/min^{-1} and a reporting limit of 100 μg. Even with an example substance having an assigned OEL of 2,000 μg/m^3 (a fairly high OEL), short sampling periods will give very high data points that always exceed the OEL. Moreover, we need to statistically analyze these data to assess the risk to the operator. Unless Bayesian methods are used, traditional analysis proves extremely difficult to implement.

It is important to note that non-specific methods are only applicable when attempting to assess the handling of single substances; that is, pure materials. If a blending operation or a granulation procedure is being assessed, then any data obtained from a non-specific gravimetric method is almost certainly going to be useless. As we have already mentioned that any dust or powder will contribute to the weight and cannot be discriminated, then we cannot utilize such methods when multiple chemical entities are present (unless the mixture itself is what is being evaluated, in which case such a method would be acceptable. This becomes a judgment call by the practicing hygienist).

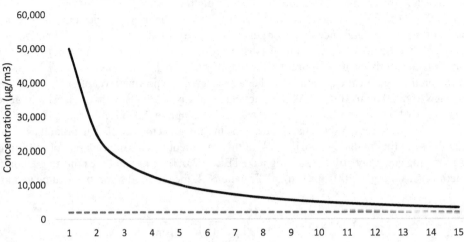

FIGURE 3.10   The limitations of non-specific sampling methods for short-term exposure sampling.

3.3.3   SAMPLING FOR PARTS OF A WHOLE

We have laid out the challenges for sampling APIs, particularly method sensitivity for substances in higher OEB tiers and particle size/dustiness. Additionally, we have discussed challenges facing hygienists forced to utilize non-specific methods, especially for short duration tasks. However, there exists a unique way to perform sampling which exploits specific methods for substances which do not have a specific method developed for them. On the surface, this may seem paradoxical. After all, how can we assess for a specific material if it doesn't have a specific IH analytical method?

Many excipients that are used exist as a salt; that is, they exist with a specific metal counterion or they contain an atom or element which can be specifically analyzed. Rather than sampling for a complete, singular substance, the practicing hygienist can sample for the material being used and assess for the unique part of the material. If the molecular formula of the substance is known (which should always be the case), the hygienist can back calculate how much of the parent material is present, and from that, the airborne concentration.

For example, assume a tablet manufacturer is using calcium phosphate dibasic as an excipient. This material is often used due to the advantageous compaction properties it exhibits in the tableting process. For the purposes of our example, let's assume the organization established an 8-hour TWA of 3,000 µg/m³ and a task-based OEL of 6,000 µg/m³ (these are both extremely high values, but we'll use them for the purposes of our example). While there is no established method for calcium phosphate dibasic, there are methods specifically for both calcium and phosphorus. In this case, the material can be analyzed for either one of the components.

We will assume the hygienist performs sampling for calcium, which has a method reporting limit of 10 µg and requires a pump flow rate of 2.0 L/min⁻¹. The task being assessed is changing of a flexible intermediate bulk container (FIBC), a task which takes an average of 14 minutes to complete. The result from this sample comes back from the lab with a level of 66 µg of calcium, which when factored in with the volume of air sampled would equate to approximately 2,300 µg/m³. However, the reported values are for calcium only, not calcium phosphate dibasic. This is why the sampling strategy is considered to be a part of a whole.

To assess the total airborne concentration of the excipient, we need to factor in the molecular weight:

$$CaHPO_4 = 136.06 \; g/mol$$

$$Ca = 40.08 \; amu$$

$$Percentage \; of \; Ca = \left(\frac{40.08}{136.06} \right) = 0.294$$

Calcium phosphate dibasic is approximately 29.4% calcium by mass. Thus, the airborne concentration reported by the lab is only 29.4% of the total mass extracted from the filter. To account for the remaining material, the reported amount must be divided by the normalizing factor, in this case 0.294:

$$Total \; CaHPO_4 = \left(\frac{66 \; \mu g}{0.294} \right) \approx 224 \; \mu g \rightarrow \frac{224 \; \mu g}{0.028 \; m^3} = 8,000 \; \mu g/m^3$$

When the normalizing factors are applied, the final airborne levels of calcium phosphate dibasic turn out to be much higher than the originally reported levels for calcium. Based on these results, which are over the task-based OEL, further work should be done to minimize airborne levels of the excipient. Similar results would be expected with analyzing for phosphorus.

There are some advantages to sampling for parts of a substance. The most important advantage is the analytical sensitivity. Significant advances in detection science have allowed scientists to take advantage in seeking ever lower levels of metals, especially in environmental sciences. For this reason, substances containing certain metal ions can be evaluated with very low levels of detection, thereby alleviating the need to use non-specific sampling methods. In our example above, the astute reader may note that using a non-specific method would have provided roughly the same result. However, if we were to fix the issue and prevent overexposures from occurring in this process, we would once again be at the mercy of method sensitivity. With gravimetric methods having a sensitivity of 100 μg, we would only be able to acquire data showing data as low as 3,500 μg/m³, whereas analyzing for calcium specifically would provide data as low 1,200 μg/m³. From a risk management point of view, minimizing uncertainty by having better data is far more preferable.

A second and notable advantage is that this practice can be applied to mixtures, not just pure materials. For instance, if an excipient blend is made in bulk which consists of starch (51%), calcium phosphate dibasic (32%), and talc (17%), then the hygienist can assess for the individual calcium component. Utilizing the same basic calculations as before, the amount of calcium phosphate dibasic could be quantified and then applied to the mixture as a whole (since the proportions of the mixture are known and presumably a constant due to manufacturing specifications). However, it should be noted that if the proportion of the substance being evaluated is rather small, then the results may not be any more accurate than non-specific sampling methods.

Just as any method has advantages, there are also disadvantages to sampling for parts of a whole. The example given above relies on metal counterions to be assessed. The number of materials which utilize such metals is relatively few. Calcium, potassium, phosphorus, tin, titanium, and zinc are some of the more applicable metals which can be assessed and utilized in such a manner. Most other metals have significant toxicity concerns and cannot be utilized in pharmaceutical manufacturing. However, a number of materials utilize sodium, magnesium, and manganese as some sort of component in their make up; unfortunately, many of the solvents used for the analytical evaluation contain relatively high levels of these metals and can often be a source of interference. Consequently, evaluating for these metals is generally not conducted.

3.3.4   ENZYME-LINKED IMMUNOSORBENT ASSAYS (ELISA)

All of the analytical methods and techniques discussed so far apply to small molecules and traditional chemical entities, yet earlier we mentioned that biopharmaceuticals have firmly established themselves in the pharmaceutical realm. Their typically proteinaceous composition more often than not precludes them from being analyzed using these analytical tools. Even so, it is not uncommon for them to have low OELs assigned to them, requiring the hygienist to perform his or her due diligence in protecting the employees handling these materials.

Analyses of biopharmaceuticals and other proteins of occupational interest by ELISA methods present an interesting alternative to the hygienist.[17] ELISA methods, while commonly utilized as a biochemical tool in research settings, is comparatively little used in the industrial hygiene realm. Such analyses have been frequently employed when evaluating for specific types of contaminants, such as enzymes[18] or allergenic animal proteins.[19] Among "traditional industrial hygienists", analysis by ELISA is not a common occurrence. Not surprisingly, most labs which cater to such hygienists do not even offer methods utilizing such an analytical method. In truth, there is a scarcity of validated industrial methods which utilize ELISA as the method of analysis, even further limiting the utility of the protocol.

Yet as was mentioned earlier, the rise to prominence of biopharmaceuticals makes the need of ELISA methods almost a universal necessity. Virtually every major pharmaceutical company has a monoclonal antibody either on the active market or in development, so it would seem only a matter of time before major industrial hygiene labs begin offering this method of analysis for various biopharmaceutical agents.

But it's worth noting that the final biopharmaceutical products are not the only substances of concern which can be evaluated by ELISA. The contemporary push for greener, climate change-friendly synthetic routes has led to many pharmaceutical routes utilize enzymatic catalysis.[20] A unique entry into the field of synthesis, enzymes are biologically active substances in that they catalyze various processes. Many enzymes are potent respiratory sensitizers and repeated occupational exposure can lead to the development of asthma. There are few occupational exposure limits given for enzymes, although the ACGIH has provided a TLV of 60 ng/m^3 for a general class of enzymes (subtilisins). In fact, this OEL is recommended even by the Association of Manufacturers and Formulators of Enzyme Products (AMFEP) for all enzymes.[21] This extremely low airborne concentration is on par with the allowable levels for HPAPIs, making it an extremely challenging target in a large-scale manufacturing setting. But with these recommended safe airborne levels, ELISA analysis offers the best methodology for the practicing hygienist to ensure worker safety.

3.4   IDENTIFYING WHERE TO SAMPLE

One of the understated aspects of crafting a valuable sampling plan is identifying where to sample. When we say, "where to sample", we mean where in manufacturing process (although physical location within a facility is important as well). Far too often the assumption is made that exposures to APIs and excipients occur only during the initial handling of the materials, and once they are "charged" or added to the manufacturing process, there are no subsequent exposures. This could not be further from the truth. There are a variety of downstream processes in a pharmaceutical operation after a substance is introduced, and any one of them could involve interaction at some level with the material. The importance of knowing the entirety of a process cannot be over emphasized. But just because multiple processes may involve the same substance, the operators of those different processes may not have the same exposure profile. This underscores the criticality of identifying appropriate similar exposure groups (see the next section for more information).

To illustrate the point, consider Figure 3.11 which shows a very simple process flow diagram (PFD) showing the flow of two solid materials to make a granulation mixture. At the bag dump station, an employee adds a material classified as an OHC 3 with an 8-hour OEL of 80 µg/m³ and a task-based OEL of 200 µg/m³. When viewed through the lens of basic compliance, the primary concern would be the 8-hour exposure for the operator who worked at the dump station, and efforts would be made to evaluate exposures to the employee. These efforts likely would have involved the operator wearing a sampling pump for an entire 8-hour shift to assess their total exposure.

But the compliance route all but ignores task-based risks. The same employee may also perform tasks at the mixer where the material is conveyed and mixed with other substances. Depending on how the material is added, there could be additional risks associated with the task, but compliance-based sampling would not identify these. Furthermore, the "other substances" being added to the mixer should also be evaluated. In Figure 3.11, these are loaded in bulk via flexible intermediate bulk container (FIBC) and conveyed to a storage silo where they remain until they are called for by the system (additional components not shown). In both processes, there are rotary valve connections which can serve as exposure points not just for the operator, but for maintenance personnel. The conveyor motors also require regular maintenance, and therefore, exposures could occur during such tasks. A final point of concern is the built-in dust control at the two stations. They are connected to a common

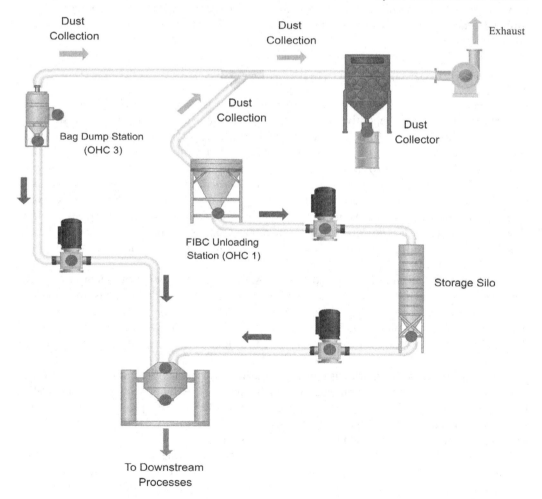

FIGURE 3.11 A process flow diagram showing general directional flow of material to a mixer. Potential exposure points are shown as red dots, and directional flow for raw materials are indicated by blue arrows. Direction of dust collection is represented by orange arrows.

cartridge dust collector where the two materials will mix in random proportions. Those tasked with changing the cartridges and also emptying the dust collector have exposure concerns.

All too often only the "initial" handling of a substance is evaluated and not any subsequent downstream (or upstream) processes. While it is true that the initial handling almost certainly presents some of the greatest risks, all of the other processes cannot be ignored. The PFD in Figure 3.11 shows but a very brief segment in the life cycle of a substance within a manufacturing facility. All aspects of material handling should be evaluated at some point, including receiving of the material, sampling by the quality department, transport of bulk material, handling and discharging of material, weighing operations, granulation, milling, formulation, tableting, and packaging. These are just a few of the major operations that go on within the pharmaceutical arena, and all should be evaluated for exposure potential.

A final point to reiterate is that when evaluating the risk of a particular task or operation, one should remove interferences or contamination from other tasks. That is, any of the above-mentioned tasks within the product lifecycle can contribute exposure to a full-shift TWA, and an employee can potentially be involved with any one of them during a work shift. So to truly evaluate each task while still evaluating the full-shift compliance, integrated sampling offers an attractive approach. Answering multiple questions in a single sampling operation is time and cost effective. The primary drawback is that the hygienist is essentially required to monitor and observe the operator the entire shift and swap samples as the tasks change, but the results obtained from such an endeavor are usually worth it.

3.5   IDENTIFYING WHO AND WHEN TO SAMPLE

Pharmaceutical employees are typically not stagnant; that is, they do not simply stay at a single station doing a single repetitive task for 8 hours. They move and rotate from one station to another, from one piece of equipment to another. Moreover, employees do not handle just a single substance while carrying out their tasks. This mixing and matching and variability provide significant complexity when crafting a sampling plan. If some employees perform the task routinely but others only occasionally do it to "help out" when needed, are these two groups at the same risk? One school of thought would say that anybody who does or can perform the task should be sampled upon, in order to have the full population. But if one of those "infrequent" users is chosen at random and they don't perform the operation, then the acquired data is heavily skewed if not altogether worthless, and everyone's time has been wasted. How then, do we identify who needs to be sampled?

The pharmaceutical industry has adopted the AIHA recommendation of placing employees into similar exposure groups, or SEGs. The creation of SEGs allows the hygienist to categorize those employees who perform the same tasks with roughly the same duration and frequency. The placement of employees into SEGs rapidly narrows the field of sampling candidates and increases the likelihood that any gathered data would be representative of the actual exposure profile for that task.

Once again, knowing the process itself can help the hygienist create SEGs. Studying the PFD can help identify not just where to sample but which equipment are used and who is trained to utilize them. But having the PFD is only the start. Prior to crafting the sampling plan, it's vital to observe the process and equipment. Direct observation provides an opportunity to talk with the operators and gain valuable insight for who to include and who to not include.

It is important to note that employees can belong to more than one SEG. In fact, it is a virtual guarantee that they will. For example, suppose we have six employees who routinely change out the FIBC in Figure 3.11. Those six operators would constitute the SEG for that task. But let's also assume that two of those operators also regularly fill in and perform tasks at the mixing tank, where the material is discharged for making the granulation blend. The same material is handled at both tasks, but the exposure profile is almost invariably different for the two tasks. They handle different quantities of the substance, different rates, different duration, and certainly different controls. Thus, the two employees who handle the same material at different stations belong to two separate SEGs.

Taking this into account, a quick glimpse at Figure 3.11 would show that for the very simple processes outlined, there exists the potential for *many* SEGs, each one requiring its own sampling plan and analysis. For those in management, they may only see that such efforts will certainly cost additional time and money and may not see the potential value in the acquired information. It is incumbent on the hygienist to convince management of the value in the sampling plan. Properly identifying risks to employees handling substances allows the organization to fully maximize their greatest asset: their employees.

An argument can be made that the two operators who work at both stations may in fact belong to a third and separate SEG. Since they perform two tasks which handle the same material, their total exposure burden is more than the operators in either SEG. While conceptually this is true, only acquired data would be able to confirm this. And again, integrated sampling would provide helpful insights as both tasks can be evaluated and the total 8-hour exposure profile as well. Further evaluation of the data using analysis of variance (ANOVA) techniques can help shed light if there is a significant variation in the two workers total exposure. If so, they could constitute their own exposure group which would require unique risk management actions. Being aware of such situations is a constant concern for hygienists in the pharmaceutical sector.

Another aspect of identifying operators for sampling is when to perform sampling. The goal of sampling is to acquire representative data of an operation to gauge an exposure profile as accurately as possible. By far the best way to do this is by random sampling. After identifying who needs sampling, deciding when to sample them should be done randomly; that is, picking random days and times to perform the sampling will give the most useful insight to the exposure profile of the task in question. For those performing integrated sampling, sampling will need to begin at the start of the employee's shift and the media changed out periodically, especially before the specific task. Scheduling the sampling with a specific employee introduces bias – he or she is prepared ahead of time and may perform the task differently, thereby skewing the data. In contrast, employees often balk when a hygienist springs upon them that sampling will be performed without their knowing. A compromise can be reached by announcing ahead of time that over a specified duration, sampling will be performed upon the employees but not informing the operators as to who and when. In this manner, they will not be as surprised, and the data should be more representative. All efforts to reduce sampling bias should be undertaken.

This presents a continual problem with consultants, or outside assistance with sampling. For most consultants, they acquire a significant amount of data, but it is usually over a very short time frame, usually within less than a week. All too often those days of data collection are contiguous (back-to-back) and the same shifts are sampled. As we mentioned above, sampling upon the same shifts and personnel can introduce significant, even if unintended, bias. Performing sampling under significant time constraints almost always leads to biased data. However, that is not to say that data acquired from consultants is not useful; on the contrary, the data can certainly paint a very vivid picture as to which tasks and processes require further evaluation.

As mentioned above, repeated sampling on the same shifts can potentially lead to sampling bias. It falls to the hygienist writing the sampling plan to understand how many workers comprise the SEG and which shifts they work. An important aspect which should always be kept in mind is off-shift workers, those who work nights and weekends. As is often the case, these shifts typically have fewer "eyes" on them providing strict oversight which can inadvertently lead to "shortcuts" in the process: not utilizing the proper PPE, disregarding certain controls, etc. While such employees would comprise the same SEG as those on conventional shifts, the data may tell a very different story regarding their exposure profile.

3.6   IDENTIFYING HOW MANY SAMPLES TO ACQUIRE AND PRESUMPTIVE COSTS

Having identified what needs to be sampled, which methods to utilize and their shortcomings, the processes which need sampling, the SEGs and when to conduct sampling, the next piece of the sampling plan involves deciding how many samples to acquire.

Perhaps no other aspect of industrial hygiene has invoked more consternation between practicing hygienists and management then deciding how many samples can be acquired. After all, there is a direct correlation between sample number and cost. Historically all that mattered was if a company was in compliance with regulatory requirements and that particular decision could be made with the fewest number of samples possible (almost always a full-shift sample, ignoring task and process variability). Fortunately, as we mentioned in Chapter 1, the pharmaceutical industry has realized that adopting a risk-based approach to protecting employees is far better policy. Consequently, they are typically able to get more samples.

Even though pharmaceutical hygienists have more freedom than some of their counterparts in other industries, restrictions still exist. In a perfect world, we would acquire dozens of samples to gauge the exposure profile as accurately and detailed as possible. Such efforts are not truly feasible, however, and choices must be made regarding the number of samples to acquire.

We have mentioned more than once the utility of integrated sampling to assess both full-shift and task-based exposures. Such a strategy is extremely useful to provide answers to both questions. However, this strategy immediately increases the number of samples to acquire, which has normally been the hurdle when formulating the sampling plan with management. If the budget is allotted enough for three samples, what has normally been meant is either three full-shift samples or three short-term samples, depending on what question is trying to be answered.

In reality though, there is no single best way to determine the number of samples beyond needing enough to perform basic statistics with the data (typically three data points). This is because the situation is entirely dependent on the operation being performed and implementation of the frequency and duration of the task(s) which can lead to exposure of the agent in question. As an example, suppose a site is making a batch of an API and there is a step in which bags of powder must be added in a bag dump station. For the production run, several bags are added, and the entire process takes about 35 minutes to complete. Importantly, once the bags are added the operator does not perform that task for the remainder of the day. Furthermore, all downstream processes involving mixtures of the material are completely enclosed, making exposure to the substance at other points in the process unlikely. In this scenario, the hygienist may decide that integrated sampling is not necessary; rather, he or she can simply acquire STEL readings to assess their peak exposure during the task and extrapolate for a full shift, assuming zero exposure for the rest of the day (this assumption is only valid if the hygienist verifies that all potential exposure points are ruled out). Such a strategy is not uncommon and still serves to obtain the needed information for assessing risk while saving considerable money and time.

But again, the path forward on determining the number of samples to acquire is truly dictated by the questions that the hygienist is trying to answer. Is peak exposure a concern to make sure respirator maximum use concentrations are not exceeded? Do multiple exposures occur throughout the day, necessitating the need for both a STEL and a full-shift evaluation? These questions truly drive the decision on how to decide on a sampling strategy. Since alignment to a risk-based model, however, hygienists in the pharmaceutical industry have largely been trying to answer both questions. The use of integrated sampling allows the hygienist to do so in an economical manner.

In practical terms, this plays out in a predictable fashion. An employee tabbed to be sampled upon for the day has a pump placed upon him or her at the start of the shift. Sampling continues until the operation in question begins. At this point, the pump is stopped, and the first sample is cataloged, followed by the implementation of a new filter. This filter is then permitted to run for a total of 15 minutes to evaluate the peak exposure, or STEL, of the process. The pump is stopped after the 15 minutes, and a new filter is used which then runs for the remainder of the task and also the shift. In this fashion, a total of three samples per person are acquired. The data will yield one STEL reading and one 8-hour TWA once averaged together. But we mentioned in Chapter 2 that typically 3 sets of data are needed for evaluation, whether peak exposures or full-shift exposures. To gather the requisite data set, the process will be completed two more times, ultimately yielding a total of nine breathing zone samples. Once acquired, the data are analyzed using the preferred method of analysis (traditional statistics, Monte Carlo, or Bayesian analysis) to compare against both the full-shift OEL and the task-based exposure limit.

By adopting the integrated sampling strategy, one can make the argument that essentially all the necessary risk-based questions are being addressed in a single sampling plan. The data provides analysis to appease regulators and auditors alike (evaluation against a full-shift OEL) but also gives insight to how controls are working surrounding the task itself (the STEL). The short-term sampling also permits evaluation of the need for respiratory protection, or if respiratory protection is already utilized that the maximum use concentration (MUC) is not exceeded for the given assigned protection factor (APF). From a risk management perspective, such information is invaluable because it informs the practicing hygienist how the currently installed controls are working. If the task-based data come back as showing an elevated exposure risk, then hygienist can begin to assess the controls to verify if they are adequate for the task and the OEB of the material.

The preceding argument should be a clear rationale for the adoption of the current strategy for integration into the sampling plan despite the increased cost the additional samples incur. Integrated sampling by its very nature triples the sampling budget, and depending on what the method and analyte happens to be, this can be significant. For example, some pharmaceutical substances carry a price tag of almost $200 per sample. If management was intending to budget approximately $600 for the project but the cost is now $1,800, they may balk at the proposal. Usually this is due to a lack of understanding of what the revised process brings in return. The data acquired from an integrated sampling approach answers multiple questions at the same time, and for organizations that are adopting risk-based approaches, such streamlined techniques are typically welcomed.

3.7 EVALUATING CONTROLS OF AN OPERATION

A good sampling plan will outline how to acquire good and useable exposure data (as they should!), but they should also include allowances for evaluating the controls which permit the exposure data; that is, the hygienist should make every effort to document all the controls which play a part in the task. By controls, we refer to the hierarchy of controls consisting of engineering controls, administrative controls, and PPE.

More on this topic will be discussed in Chapters 5 and 6, but briefly engineering controls are equipment and facilities which isolate, reduce, move, or otherwise keep airborne contaminants away from operators. These can include isolators, gloveboxes, general room ventilation, local exhaust ventilation, hoods, and other pieces of equipment. During sampling efforts, all engineering controls currently in place should be documented and their activity measured. Simply stated, if it is designed to move air, then air flow should be measured. If it is designed as a barrier, the effectiveness of that barrier should be assessed (exposure data itself will give an indication for this). Measuring current engineering controls and then correlating exposure data to the performance of said equipment allow the hygienist to correlate exposure to control performance, thereby being able to recommend improvements and upgrades.

Likewise, administrative controls need to be assessed. These include work practices, training, and other "intangible" aspects to the job being performed. Such evaluations can often take quite a bit of time, but parts of this are already conducted when evaluating members of the SEG. It is typically only a small additional step to further look at any SOPs to see if they are accurately followed or training records to see how in-depth they are or their frequency of occurrence. Again, exposure data can be tied to administrative controls and give an avenue for improvement.

Finally, the PPE being used in a task should also be evaluated. It is a frequent occurrence for PPE to be chosen almost at random and not based on the physical or chemical properties of the materials being handled. Likewise, respiratory protection is often assigned without knowing if the assigned protection factor is appropriate based on the exposure profile (this is a very common occurrence in industries), and it is critical to evaluate fit test records of the SEG members. The proper PPE use, which should be documented in a well-crafted SOP, will certainly help minimize exposures, but PPE should be used only as a last resort. Furthermore, not all employees are trained on appropriate donning and doffing protocols, and these practices become extremely important when handling more potent compounds. In short, the sampling plan may not necessarily spell out that controls need to be evaluated during sampling sessions, but they absolutely should be as part of a routine industrial hygiene sampling effort.

3.8 SUMMARY

In this chapter, we delved into the facets of the industrial hygiene exposure assessment as they relate to the risk assessment paradigm. In contrast to traditional exposure assessment for general populations, such as those conducted by the US EPA and the WHO, industrial hygiene exposure assessment is able to directly and quantitatively measure employee exposures in the workplace. The advantage of being able to do so is tremendously useful. But to evaluate exposures, the hygienist must first acquire representative and useful data. To obtain the most representative data possible regarding the tasks performed and exposure profiles, a detailed sampling plan must first be crafted.

The sampling plan answers the questions of who, what, when, where, and how as they relate to employee exposures: who to sample upon (the SEG of interest), what are we sampling for, when will sampling take place, where (in the process or what tasks to evaluate) will sampling take place, and how will we analyze the substance (analytical method). The sampling plan also lays out the anticipated number of samples and presumed costs based on the laboratory scheduling fees. Once approved by all parties, there should be no surprises and execution of the sampling plan can be performed by any practicing hygienist. Most importantly, the sampling plan directly provides valuable input as it relates the risk management process. A well-crafted sampling plan provides the third piece of the risk assessment puzzle for the risk management team: exposure data. This aspect fits directly into the ISO 31000 mantra while maintaining the spirit of the NRC risk assessment methodology.

NOTES

1 Risk Assessment Forum. (2019, October). *Guidelines for Human Exposure Assessment.* (EPA/100/B-19/001). US Environmental Protection Agency.
2 Mulhausen, J. "Faulty Judgment", Synergist, November 2021.
3 Logan, P., Ramachandran, G., Mulhausen, J., and Hewett, P. (2009). Occupational exposure decisions: Can limited data interpretation training help improve accuracy? *Annals of Occupational Hygiene*, 53(4), 311–324.
4 Vadali, M., Ramachandran, G., Mulhausen, J. R., and Banerjee, S. (2012). Effect of training on exposure judgment accuracy of industrial hygienists. *Journal of Occupational and Environmental Hygiene*, 9(4), 242–256.
5 Rowe, R. C., Sheskey, P. J., and Quinn, M. E. (2009). *Handbook of Pharmaceutical Excipients*, 6th Ed.
6 Perez-Ibarbia, L., Majdanski, T., Schubert, S., Windhab, N., and Schubert, U. S. (2016). Safety and regulatory review of dyes commonly used as excipients in pharmaceutical and nutraceutical applications. *European Journal of Pharmaceutical Sciences*, 93, 264–273.
7 Swarbrick, J., 2006. Coloring agents for use in pharmaceuticals. In: Schoneker, D. R. (Ed.). *Encyclopedia of Pharmaceutical Technology*, pp. 648–670.
8 de la Torre, B. G. and Albericio, F. (2021). The pharmaceutical industry in 2020. An analysis of FDA drug approvals from the perspective of molecules. *Molecules*, 26(3), 627.

9   https://www.fda.gov/about-fda/center-biologics-evaluation-and-research-cber/what-are-biologics-questions-and-answers. Accessed July 14, 2022.

10 https://www.fiercepharma.com/special-report/top-20-drugs-by-2020-sales. Accessed December 2, 2022.

11 Gilleskie, G., Rutter, C., and McCuen, B. (2021). *Biopharmaceutical Manufacturing: Principles, Processes, and Practices.* Walter de Gruyter GmbH & Co KG.

12 Graham, J. C., Hillegass, J., and Schulze, G. (2020). Considerations for setting occupational exposure limits for novel pharmaceutical modalities. *Regulatory Toxicology and Pharmacology, 118,* 104813.

13 Olivier Jr, K. J., and Hurvitz, S. A. (Eds.). (2016). *Antibody-Drug Conjugates: Fundamentals, Drug Development, and Clinical Outcomes to Target Cancer.* John Wiley & Sons.

14 *Chemical Linkers in Antibody-Drug Conjugates (ADCs).* van Delft, F.; Lambert, J. M. (Eds.). Royal Society of Chemistry. © 2022.

15 "Cytotoxic Payloads for Antibody-Drug Conjugates" Thurston, D. E.; Jackson, P. J. M. Royal Society of Chemistry. 2019.

16 ISPE Good Practice Guide. "Assessing the Particulate Containment Performance of Pharmaceutical Equipment, 2nd Ed." 2012.

17 Premjeet, S., et al. (2011). Enzyme-linked immuno-sorbent assay (ELISA), basics and it's application: A comprehensive review. *Journal of Pharmacy Research, 4*(12), 4581–4583.

18 1) Houba, R., Van Run, P., Heederik, D., and Doekes, G. (1996). Wheat antigen exposure assessment for epidemiological studies in bakeries using personal dust sampling and inhibition ELISA. *Clinical and Experimental Allergy, 26*(2), 154–163. 2) Houba, R., Van Run, P., Doekes, G., Heederik, D., and Spithoven, J. (1997). Airborne levels of α-amylase allergens in bakeries. *Journal of Allergy and Clinical Immunology, 99*(3), 286–292.

19 1) Harrison, D. J. (2001). Controlling exposures to laboratory animal allergens. *ILAR Journal, 42*(1), 17–36. 2) Hollander, A., Van Run, P., Spithoven, J., Heederik, D., & Doekes, G. (1997). Exposure of laboratory animal workers to airborne rat and mouse urinary allergens. *Clinical & Experimental Allergy, 27*(6), 617–626.

20 Tao, J., & Xu, J. H. (2009). Biocatalysis in development of green pharmaceutical processes. *Current Opinion in Chemical Biology, 13*(1), 43–50.

21 "Guide to the safe handling of industrial enzyme preparations". AMFEP guidance document. https://amfep.org/_library/_files/amfep-guide-on-safe-handling-of-enzymes-updated-in-2013.pdf.

4 Industrial Hygiene Risk Assessment
Risk Characterization

4.1 INTRODUCTION

The definition and concept of the risk characterization have somewhat changed over the years. The original 1983 NRC statement on risk characterization was "…the final expressions of risk derived in this step will be used by the regulatory decision-maker when health risks are weighed against other societal costs and benefits to determine an appropriate action".[1] No further advice or guidance was provided for this step. Ultimately the NRC amended their stance on risk characterization and in 1996 revised the definition to be "…a synthesis and summary of information about a potentially hazardous situation that addresses the needs and interests of decision makers and of interested and affected parties. Risk characterization is a prelude to decision making and depends on an iterative, analytic-deliberative process".[2] The EPA has also modeled their interpretation of risk characterization around the NRC definition, stating "…the risk characterization integrates information from the preceding components of the risk assessment and synthesizes an overall conclusion about risk that is complete, informative, and useful for decision makers".[3]

For risk assessors and industrial hygienists performing a human health risk assessment, the final step is the risk characterization (Figure 4.1). As with the previous steps of the risk assessment, the risk characterization fits nicely into the ISO 31000 workflow and encompasses the steps of "risk analysis" and "risk evaluation". In Chapter 1, we detailed how "risk analysis" referred to the either quantitative or qualitative description if a risk exists, and the "risk evaluation" is the step in which the organization decides how they will treat said risk.

In Chapters 2 and 3, we detailed the three key input components to the NRC risk assessment paradigm: hazard identification, dose-response assessment, and exposure assessment. These three steps make tangible sense as they depend on the acquisition of evidence, of scientific data to dictate to the risk assessors the nature of the chemical hazards and just how much employees are being exposed to on a routine basis. The final step, risk characterization, is by its very nature vague and uncertain. If the preceding steps entail the acquisition of information and data, the risk characterization step attempts to answer the question "how do we interpret this data?"

Importantly, we want to interpret the data for the *entire exposure group*, not just a single individual. Interpretations of exposure data for a single individual are relatively simple and can be conducted following basic compliance calculations such as those found in the "NIOSH Occupational Exposure Sampling Strategy Manual".[4] Perhaps more simply, many hygienists will simply look at the TWA data for a single individual and if they are below the established OEL, will state that the employee is "under the exposure limit" and therefore in compliance (however, this is NOT best practice and the practicing hygienist should utilize the NIOSH equations for compliance calculations). But such evaluations apply only to the singular individual, not to the entire SEG. When we speak of risk characterization in industrial hygiene, we are really speaking to the overall risk to the SEG, and given the fact that no two employees perform a particular task the *exact* same way with the *exact* same exposure profile, there needs to be a different approach.

In this chapter, we will focus on how representative industrial hygiene data are analyzed within the pharmaceutical industry and how the industry uses those analyses to classify and judge the risk to the

DOI: 10.1201/9781003273455-4

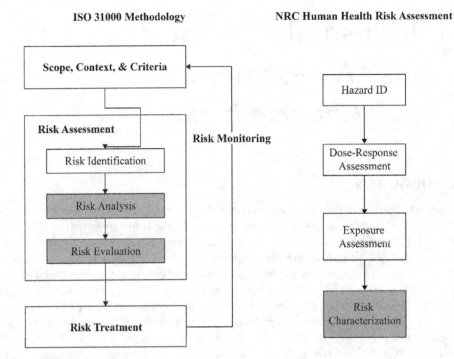

FIGURE 4.1 The ISO 31000 risk management workflow and how the NRC risk assessment paradigm fits into it for human health risk assessment. The NRC risk characterization is identical to the ISO 31000 risk analysis and risk evaluation steps.

operators. We will see some of the more common means of analyses and demonstrate their shortcomings and modern work arounds to these issues. We will also delve into how these data, representative of an entire SEG, are then labeled so they can be subsequently mitigated via risk treatment.

4.2 ESTABLISHING CRITERIA FOR SEG RISK CHARACTERIZATION

4.2.1 THE 95TH PERCENTILE

Before going into detail on how industrial hygiene related risks are characterized for a group of workers, it is imperative to understand a basic risk management tenant as it applies to exposure profiles. As industrial hygienists, we want to ensure that virtually all employee exposures are below the allowed OEL for a given material. But from a risk management perspective, what is the defined criteria that is deemed acceptable for employee exposures?

The field of industrial hygiene has independently developed an answer to this issue without any prior knowledge or formal application to the modern field of risk management. For any SEG, there is going to be variations among the data, leading to an exposure profile or exposure distribution. Virtually all IH data are lognormally distributed, as shown in Figure 4.2. The American Industrial Hygiene Association (AIHA) and other authoritative IH bodies have formally recommended the use of the 95th percentile as the governing line for acceptance criteria of an exposure profile. By evaluating where the 95th percentile of an exposure profile falls within context of the OEL of a substance, hygienists gain a degree of confidence as to how well the *entire SEG* is protected, not just a single individual. By doing so takes the variability between worker exposures into account. In other words, if the 95th percentile of an exposure profile is below the OEL, the hygienist can say that 95% of all exposures (employees) in the SEG are below the OEL. From a risk management perspective, we are saying that it is an acceptable risk for 5% of exposures to potentially be above the OEL, since we are protective of most of the employees.

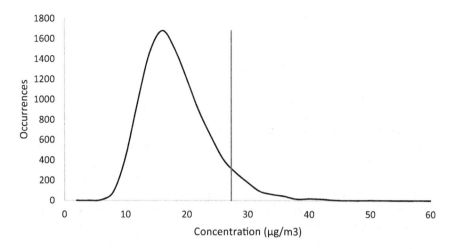

FIGURE 4.2 A typical log-normal distribution with the 95th percentile (upper tail) highlighted. If the 95th percentile for an SEG's exposure profile is below the OEL, then the task is considered acceptable.

These concepts are important to note since they inherently encompass the mantra of risk management. By using the 95th percentile as our acceptability criteria, we are not trying to outright eliminate risk to employees. In contrast, we are saying that there is an acceptable amount of risk that an organization is willing to accept, and to obtain those levels of risk requires risk management. This is the very reason that industrial hygiene fits so well into the ISO 31000 risk management paradigm: there is an acceptable level of employee exposure (risk), and as an organization, we must manage the controls (variables) which contribute to the risk.

4.2.2 AIHA EXPOSURE CATEGORIES

Historically when evaluating an exposure profile for an SEG, hygienists were concerned as to whether the 95th percentile was simply above or below the OEL. If the 95th percentile was below the OEL, the process was deemed acceptable and no further action was recommended or warranted; however, if the 95th percentile was above the OEL, then changes were recommended for the operation, which could include the use of PPE such as respiratory protection. In this manner, there are only two options: acceptable and unacceptable. There was no designation for degrees of acceptability or unacceptability. For instance, if an assessment showed the 95th percentile of an exposure profile to be 97 ppm and the OEL was set at 100 ppm, the process would be given a positive review and everything would be "green-lighted", so to speak. To account for this apparent disparity, the AIHA laid out a systematic process to establish categories of exposure. The Exposure Assessment Strategies Committee (EASC) codified a series of guiding principles for industrial hygienists so there could be a simple and effective means of categorizing exposures. The first publication date was in 1991 and has constantly been revised in the years since.

The AIHA EASC recommended the use of a 5-category system which was delineated by percentages of a given OEL. These categories were designated 0 through 4, with 0 being the lowest exposure category and represented exposures wherein the 95th percentile was less than 1% of the desired OEL. A Category 1 exposure was defined as those whose 95th percentile was less than 10% of the limit. In a similar fashion, Categories 2, 3, and 4 were defined as those exposures that were less than 50% of the limit, less than 100% of the limit (at the OEL), and in excess of the limit, respectively. The ease and utility of the system allowed for application to any exposure limit and were not relegated to any type of exposure. Along with the ease of use, the EASC also recommended a series of controls to be implemented based on the category of exposure, adding yet additional value to the system. A final point is that while AIHA recommends the 95th percentile as the decision statistic, it is not a requirement. Organizations can utilize varying decision criteria, such as the 90th percentile or the 99th percentile, but this must be decided upon by the organization and agreed upon by all parties. The remainder of this chapter will focus on the 95th percentile as being the primary criteria for determining risk.

The AIHA EASC exposure categorization scheme fits well into the pharmaceutical and consumer healthcare industries. Its incorporation completely negated the use of subjective qualitative statements for identifying a likelihood category (more on this later). In this manner, the risk assessment process moved from the qualitative methodology into a more quantitative arena. Indeed, the area of quantitative safety risk assessment as a general practice is a blossoming field but has yet to achieve widespread acceptance. This may be due in part to the fact that most individuals have a difficult time applying quantitative methods to assessing probabilities of occurrence.[5] Regardless, the AIHA EASC categories provide a useful way to differentiate between different exposure profiles and treat them accordingly in a risk management system.

4.3　INDUSTRIAL HYGIENE RISK CHARACTERIZATION

4.3.1　Qualitative Characterization: Risk Matrices

Up to this point in this book, we have dealt with and defined risk as an undesired event that can impact an organization. From an industrial hygiene perspective, the undesired event is the overexposure of employees to harmful substances. However, in Chapter 1, we also stated that another means of defining risk is the product of an identifiable hazard and its likelihood of occurrence (Risk = Hazard × Likelihood). These two definitions and approaches may seem like two statements reiterating the same thing in different words, but the ramifications of their approaches are vastly different. The approach described thus far, focusing on the first definition of risk, has been extremely quantitative in nature: we utilized experimental toxicology data to derive OELs (hazard ID and dose-response) and sampling data (exposure assessment) to quantitate employees' exposure levels. Everything described thus far is very technical and time consuming, approaches most organizations have historically not wanted to adopt.

From a wholistic view, almost every organization (including pharmaceutical companies) assesses general risks in a qualitative fashion, not quantitative; that is, they assess impact of a hazard and the likelihood of that hazard occurring in a subjective manner. For instance, hazard impact can be assessed as "negligible", "marginal", "significant", or "catastrophic". Likewise, the likelihood fixture can be assessed as "rare", "infrequent", "possible", or "likely". In order to align with the frequently utilized "risk equation", risk assessors place the two variables along two axes, forming a risk matrix. Wherever the two identified variables cross on the table is the identified and "characterized" risk level for that risk. A hypothetical 5 × 5 risk matrix is shown in Table 4.1.

TABLE 4.1
A Qualitative 5 × 5 Risk Matrix

Likelihood	Hazard Impact				
	Negligible	Marginal	Significant	Critical	Catastrophic
Rare	Low	Low	Low	Low	Medium
Infrequent	Low	Low	Medium	Medium	High
Occasional	Low	Medium	Medium	High	High
Probable	Low	Medium	High	High	High
Certain	Medium	High	High	High	High

The appeal of risk matrices is undeniable. They are simple, efficient, and easily aid risk managers in prioritizing risks based on their direct characterization. For example, it is intuitive that a "high" risk should be treated before a "medium" or even a "low" risk. The application of a color-coded system for the levels of risk provides a rapid risk conveyance system to not only risk assessors themselves but other stakeholders as well. A typical color scheme is the RAG (red, amber, green) approach: red is used to highlight serious or unacceptable risks that need to be mitigated immediately; amber risks are those which need to be rectified but are not an immediate priority; green risks are considered acceptable. All risks from the same category are given the same color. Another significant appeal is that such risk assessments can be performed extremely rapidly and applied to virtually any hazard in any field or application. A risk assessor can quite literally walk through any area of an active worksite and rapidly assess/characterize any number of hazards quickly and efficiently. Documentation can also be provided to strengthen the risk assessment.

The aforementioned risk matrix is entirely qualitative, and therefore subjective, in nature. While there is certainly some logic in the ability to prioritize a "high" risk from a "medium" risk, one can glean from the matrix itself that there are multiple entries for each risk level. Should these all be treated the same and given the same number of resources to mitigate? It is not possible to further subdivide the "characterized" risks from the risk matrix. Furthermore, such characterizations do not strictly meet the risk equation requirements since words can't be multiplied together to give a product. This single, albeit profound, shortcoming of purely qualitative risk characterizations via risk matrices led to an "improvement" that is often referred to as semi-quantitative risk characterization via the introduction of risk scores.

The implementation of risk scoring systems is rather straightforward. Lower-level categories, of both likelihood and hazard impact, are given lower "values". These typically begin at 1 and incrementally increase by a single integer per risk factor. When applied to our previous 5×5 risk matrix, we get Table 4.2.

The output of Table 4.3 is considerably different than in Table 4.2. By having numerical values assigned to each risk factor, the risk categories are rapidly and easily subdivided. The values in each cell are obtained by matching the appropriate "hazard impact" level with the corresponding "likelihood" level and multiplying their values. Importantly, the color-coded areas remain the same between the two tables, establishing the boundaries of what is numerically acceptable for risk scores. In this hypothetical scenario, risk scores < 5 are considered "green" or acceptable; those with scores ≥ 5 but < 10 are considered "amber", whereas scores of 10 or greater are characterized as serious risks. This is part of the risk criteria that is often established prior to a risk management program being implemented. To help facilitate such risk assessments and characterizations, organizations go to great lengths to define each of the characterizing risk factors, that is "likelihood" and "hazard impact". Indeed, there is no single authoritative standard which defines what these risk factors should be; rather, it is up to each and every organization to craft their own definitions of these terms. In essence, this begins to develop an organization's "risk appetite", or what it considers to the boundaries of acceptable versus unacceptable risk.

TABLE 4.2
A Semi-Quantitative 5 × 5 Risk Matrix

	Hazard Impact				
Likelihood	Negligible (1)	Marginal (2)	Significant (3)	Critical (4)	Catastrophic (5)
Rare (1)	1	2	3	4	5
Infrequent (2)	2	4	6	8	10
Occasional (3)	3	6	9	12	15
Probable (4)	4	8	12	16	20
Certain (5)	5	10	15	20	25

TABLE 4.3

A 5 × 5 Risk Matrix Utilizing AIHA Exposure Categories and OEBs as Criteria

		Occupational Exposure Band (OEB)				
		1	2	3	4	5
	1	1	2	3	4	5
Exposure Category	2	2	4	6	8	10
	3	3	6	9	12	15
	4	4	8	12	16	20
	5	5	10	15	20	25

While there are no standard definitions for each risk factor, some generalized example definitions for the hazard risk factors are:

- **Negligible (1)** – injuries not requiring first aid treatment; incidents resulting in less than $1,000 in property damage;
- **Marginal (2)** – injuries requiring first aid treatment, but not requiring restricted work or missed time; incidents resulting in property damage losses between $1,000 and $25,000;
- **Significant (3)** – injuries requiring medical treatment and resulting in restricted work or missed time; OSHA recordable injuries; incidents resulting in property damage losses between $25,000 and $50,000;
- **Critical (4)** – disabling injuries or illness to workers; OSHA reportable injuries; incidents resulting in property damage losses between $50,000 and $200,000;
- **Catastrophic (5)** – one or more fatalities; incidents resulting in property damage losses in excess of $200,000.

Likewise, some example definitions for the likelihood risk factors could include:

- **Rare (1)** – not likely to occur; has not occurred within the last calendar year;
- **Infrequent (2)** – has a possibility to occur; has occurred within the last 6 months;
- **Occasional (3)** – has been known to occur; has occurred at least once within the last 3 months or at least twice within the last 6 months;
- **Probable (4)** – is likely to occur; has occurred at least once within the last month or at least twice within the last 3 months;
- **Certain (5)** – is very likely to occur; has occurred more than once in the last month.

Note that the definitions presented here are very safety-centric; however, they can easily be adapted to include any field or department within an organization. These definitions are very generalized examples but showcase the broad utility to which they are applied. They establish frequency boundaries by which risk assessors can gauge occurrence and levels of impact relating to severity and types of injuries or property damage. The exact boundaries defining the acceptable levels of risk for each risk factor are decided prior to the implementation of the risk management program.

The use of semi-quantitative risk matrices has taken the risk assessment field by storm. Almost every company and every industry utilize such a tool to "measure" the risk to their organization. Perhaps more importantly, the system has formed the foundation of the entire risk management program, encompassing not just the realm of safety but every aspect of an organization. We mentioned in Chapter 1 that within the pharmaceutical industry, quality is heavily regulated and requires a thorough treatment of risk in order to be properly implemented. Risks to the quality program are typically assessed via semi-quantitative risk matrices, just as risks to cybersecurity, supply chain, inventory levels, project management, and even safety are evaluated the same way. In this manner, they are assessed on an equal playing field and can be similarly tracked and prioritized in a risk management system (RMS). Such a system is often closely watched by higher-ups in companies so as to keep tabs on how the overall risk to the company is being managed. It also naturally gives rise to much vaunted metrics that enable groups and departments to quickly showcase how well (or not so well) they are performing by not only identifying risks but managing their mitigation in accordance with company policy.

In this manner, the use of semi-quantitative risk analysis "fits perfectly" into the ISO 31000 framework. There are risk criteria that need to be established beforehand, methods for assessing risk and communicating the results to stakeholders. There is a convenient system in place that allows risk managers to prioritize risks via numerical risk scores, allowing for convenient resource allocation for their mitigation. Additionally, the risk matrix can be customized to fit the needs of the organization: that is, instead of a 5×5 matrix, perhaps an organization determines that it requires a 5×4 matrix, or perhaps a 7×5 matrix. The output is essentially dictated by how they define their risk criteria, but the flexibility is a significant draw for most companies. And finally, with having a single system by which all organizational risks are evaluated gives an opportunity for management and executives to oversee and evaluate the risk program by having a chance for continuous improvement.

4.3.2 Difficulties with Applying Risk Matrices to Industrial Hygiene Risk Characterization and Other Significant Drawbacks

We previously stated that the use of risk matrices and semi-quantitative risk characterization has become the standard approach within virtually every industry and has been applied to almost every field. The use of a single risk characterization tool with defined risk parameters that can be quickly utilized by every group in a company allows for all risks to be identified, measured, and prioritized with equal weight. However, industrial hygienists have continually found the use of risk matrices to be difficult when applied to their trade.

Toward the beginning of this chapter, we stated that the field of industrial hygiene established its own risk criteria for an SEG to be where the 95th percentile of the exposure distribution falls relative to the accompanying OEL. This is a universally accepted means of protecting virtually all workers within the industrial hygiene realm. But the difficulty arises when trying to apply the use of the 95th percentile to the example definitions provided for the likelihood and hazard impact risk factors. How does one assess whether the exposure has an impact that is considered "marginal" or "significant" based on the given definitions? Furthermore, how does one assess the likelihood of an exposure profile using the definitions for "infrequent" or "probable"? Such assertions make more sense if we are assessing a singular endpoint, such as overexposure; in this format, the risk assessors would truly be asking themselves "what is the likelihood of employee overexposure in a given process?" Even so, how does one align the concept of an overexposure to the hazard impact criteria? The definition itself does not readily accommodate the rationale. Furthermore, the likelihood definitions do not align with the reality of many chemical handling processes. It is not uncommon for a process to occur daily, or perhaps several times a day. Even if the definitions are altered to align the IH observations with qualitative assessment paradigm, the risk characterization still does not align with the overall impact as dictated by the risk matrix (is a potential overexposure equivalent to an incident resulting in multiple fatalities? Or an incident resulting in $500,000 of property damage? The overexposures are dangerous, certainly, but probably not as extreme as that).

Another not insignificant issue with risk matrices and industrial hygiene is the subjective nature of qualitative assessments. In Chapter 3, we discussed in detail the pitfalls that plague industrial hygienists when they attempt to assess exposures in qualitative terms. Human beings simply are not very good at gauging exposures based on visual observation alone. Even when knowledge of the process and chemical properties are factored in, humans tend to underestimate exposures, often by significant margins. This is an inherent form of bias. Risk assessors try their best to minimize bias, but when assessments are performed qualitatively (that is, without standardized measurement tools), bias will always be introduced and factored in, even subconsciously. Adding additional details and layers to the definitions does not truly aid in the assessment, either. What one person may classify as a "critical" hazard impact another may determine to be a "marginal" impact; likewise, two assessors can come to different conclusions for the probability assignment as well. The inherent subjective nature of the analysis gives rise to often wildly inconsistent risk characterizations for industrial hygiene processes.

Some risk managers have altered the risk matrix to be industrial hygiene centric in an effort to address these shortcomings. The "likelihood" risk factor is removed altogether and is replaced with the AIHA exposure categories. A notable change is that the exposure categories on the risk matrix are assigned values of 1–5, whereas AIHA assigned them values of 0–4. This change was intentional so when the "Risk=Hazard×Likelihood" equation is utilized, multiplying values by zero is avoided. The "hazard impact" risk factors are also replaced, but with the applicable occupational exposure band (OEB) to which a substance belongs. Recall from Chapter 2 that many substances may not have a definitive OEL but may be placed into an OEB. These bands are often designated as numerical values and serve as the convenient multiplier along the adjacent risk matrix axis. Importantly, since each organization has its own unique number of exposure bands, this ultimately changes the resulting layout of the risk matrix. For our purposes, we will assume the use of a 5 band OEB system. The end result is a 5×5 matrix as seen in Table 4.3.

The newly redefined risk matrix is now applicable to industrial hygiene but identical in setup to that found in Table 4.2. We have an axis that accounts for the "hazard" of the materials being handled in an operation (the OEB) and an axis that accounts for where the 95th percentile of an exposure profile is likely to be. In this regard, a hygienist (or any risk assessor) can walk through an area and characterize a process by understanding what is being used and where the operator's exposure profile may be based on the observation. Importantly, these characterizations are done using a subjective, semi-qualitative assessment of the exposure as described previously. Such a tool is a common practice in organizations that have many processes that require "prioritization" for sampling. This approach saves time and money, again providing an avenue for productivity by enabling departments to quickly perform risk assessments. Finally, since the risk matrices in Tables 4.2 and 4.3 are identical in nature and setup, it allegedly allows for direct comparison of the IH-related risks to those characterized for any other process.

Unfortunately, while this common compromise may appease most risk managers, it is still rife with problems regarding industrial hygiene risk characterization. The first significant issue lies with the risk matrix itself, Table 4.3. There are some notable logic gaps associated with it, relating predominantly to the risk outputs. For instance, we had (arbitrarily) assigned risks with values from 5 to 9 as "amber" risks; that is, they were not considered major risks but still required some sort of action plan to address the risk. But a deeper look into the numbers reveals a problem. There are two risks with the value of "5", but one is listed as an Exposure Category 4 (one in which the 95th percentile is over the assigned OEL) and the other an Exposure Category 1 (one in which the 95th percentile is identified as being less than 1% of the assigned OEL). The flaws should be obvious: A virtually guaranteed overexposure is given the same amount of weight as one which has an almost certainty of being less than 1% (also completely ignoring the fact that an overexposure would not be listed as an amber risk, but instead a red risk requiring immediate action). The source of the error lies with the OEB Categorization axis. The lower band category results in a lower multiplier, whereas higher bands have higher multipliers. While it makes sense mathematically, pragmatically it causes issues. Simply because we are using a substance that would be listed in an OHC 5 band doesn't necessarily mean we are automatically at grave risk. If the exposure level is assumed to be a Category 1 exposure (via subjective analysis) at less than 1% of the OEL, then there is very little risk to the operators. Such outputs should not be classified as higher risk simply because of the nature of the compound (the nature of the compound does, however, dictate the required controls for handling it). This type of issue is not relegated to the risk categories of "5", either. A look at the whole risk matrix shows several instances of this sort of logical shortcoming with numbers that repeat within the table. It is important to remember that the use of the AIHA exposure categories has correlations to actual defined exposure boundaries, and so their use has physical ramifications.

A similar issue comes to light with the prioritization of the risks. It would be widely agreed upon that "red" risks must be treated before the "amber" risks, and that the higher valued "red" risks get treated first. But many of these so-called risks are below the action level, which is less than an Exposure Category 3. Such exposures are normally considered acceptable by the AIHA, but the risk matrix would require additional actions to be carried out. Such actions require man hours for investigation and potential capital investment to fix a non-existent issue. In these situations, additional work is being created and resources are being wasted. This is a recurring issue in many organizations, and it can often be traced back to poor risk categorization as a result of the scheme being used.

Most of these issues can be "solved" by using a different risk matrix, one with more available values such as a 6×5 table or a 7×6 table. Larger risk matrices offer more risk values which can give greater flexibility to characterize risk without the issues which plague the 5×5 table. But this introduces yet another dilemma. We have already established that the Exposure Categories are based on the recommendations by the AIHA, which means they have been endorsed by an authoritative body which has extensive knowledge in the field of exposure assessment. Furthermore, their recommendations on how to divvy up the exposure categories are not arbitrary: they are largely derived from what OSHA requires in many standards. Exposure profiles that are more than 50% of the OEL but less than the OEL is the Action Level, a category in which the employees are not necessarily over-exposed, but which requires action to lower the airborne levels anyway. And exposures in excess of the OEL are clear OSHA violations. Any attempt by an organization to stray from these exposure ratings would have significant outcomes. For instance, it could create more work to make the Exposure Categories fit the risk matrix (such as by creating a new Exposure Category in which the 95th percentile was greater than 75% of the OEL but less than the OEL), but this unnecessarily makes the exposure profiling more complicated. Furthermore, it does not align with OSHA's exposure nomenclature.

The problems of trying to force industrial hygiene exposure risks into a convenient risk matrix for the sake of a risk management system continue to plague practicing hygienists to this day. All too often a singular risk management system is used which contains all known risks to an organization from every department, division, and group. While the idea of having a single location to track and prioritize all risks makes sense initially, it does not take long to see the system become jumbled because not all fields will assess their risks the same way. Attempting to derive a numerical integer which is unitless to associate a risk level to it is almost meaningless. This presents another problem with this method of assessing risk. The assigned risk numbers have no actual value. The numbers themselves are what are referred to as "ordinal data", which are data with a number assigned to it, allowing for ranking, but the actual value among them cannot be quantified.[6] This is because the numbers hold no actual units. In his book *The Failure of Risk Management*, author Doug Hubbard writes extensively about how risk managers continue to inadequately manage the risks to their organizations in part because the assessment of risks using ordinal scales such as risk matrices leads to inaccurate or meaningless conclusions.[7] Those groups which use more quantitative methods have far greater success at lowering their risks because of their ability to measure their risks rather than estimating along ordinal scales. With industrial hygiene inherently being a very quantitative field by its very nature, forcing the results of assessments into an ordinal scale alters the risk output. The final risk score that gets cataloged in the system loses much of the original meaning and leads to inaccurate conclusions, as we saw in the preceding paragraphs.

In reality, the issues associated with industrial hygiene and risk matrices are not unique to the field of industrial hygiene. A growing body of research is beginning to show that risk matrices themselves are unsuitable to assessing any type of risk.[8] One significant drawback is that risk matrices suffer from "poor resolution"; that is, the use of semi-quantitative assignments often provide identical qualitative risks (such as "green" or "amber") to "quantitatively" very different risks. A risk matrix can also improperly assign a risk category to a risk. In other words, something can be of very low risk yet still be classified as a significant risk. We demonstrated this issue in Table 4.3 with the example of the risk ratings of 5 in which a higher tier substance (OEB 5) was assigned an exposure level of 1, which is less than 1% of the OEL. The significant misalignment on the significance of the risk is evidence of the inadequacy of risk matrices for evaluating risk. This, in turn, leads to a third issue with risk matrices, which is improper allocation of resources. The improperly categorized risk would require the risk management team to put together a project to mitigate the risk, costing the institution time, money, and other resources. These are resources which could be put to better use elsewhere, perhaps toward other aspects of the company which do pose a genuine risk that requires treatment, but in this particular example the resources would be wasted on fixing a non-existent risk.

4.3.3   QUANTITATIVE CHARACTERIZATION: IDENTIFYING THE 95TH PERCENTILE

The previous section was spent describing a commonly used qualitative approach for characterizing industrial hygiene risks using generalized, subjective risk criteria which can be applied via direct observation. The approach is rapid, costs next to nothing to implement, and circumvents the need for sampling, which is often viewed as a blessing by those in management. Yet problems associated with the use of a qualitative or semi-quantitative risk matrix-based system are numerous and are the reason that most industrial hygienists within the pharmaceutical sector have turned to more direct, data-driven methods to assess risk for the 95th percentile of their SEGs which are more accurate. Moreover, they are methods which rely upon sampling, not subjective inference of a process.

But exactly how do we identify the 95th percentile for our SEG? As we mentioned previously, most hygienists take multiple samples from a given task or full-shift and simply looking to see whether the laboratory data are below the limit is not an applicable means of assessing the exposure distribution. Within the pharmaceutical industry, the classical means of analyzing exposure data has been via traditional statistics; however, in the last decade or so, the industry has made a significant change and begun using Bayesian statistics to evaluate its data sets. We will look at both approaches, beginning with traditional statistics.

4.3.3.1   Traditional Statistics

For as long as industrial hygiene programs have been in existence, there has been an ongoing "battle" between hygienists obtaining a proper number of samples to determine an exposure profile and senior business management seeing the value in paying for more samples than was "legally necessary". This usually results in the hygienists taking the fewest number of samples possible. Unfortunately, this situation regularly continues to this day. Given this shortcoming, industrial hygienists turned to the next best answer for their programs: statistics. The power of statistics cannot beoverstated, as it allows users to make informed decisions with minimal input. However, it's imperative to understand that there is significant uncertainty surrounding the statistical estimates, as we will see later.

We mentioned earlier how sampling data are distributed. Almost universally, samples will present data that is lognormally distributed; that is, if the logarithm of the data points were taken, those log-transformed data points would be normally distributed. This is important to note because it necessitates a few additional steps when analyzing data.

The steps for analyzing log-normal data in industrial hygiene and the shortcomings surrounding uncertainty are best explained with an example. Suppose a hygienist in a consumer healthcare plant is concerned about employee exposures to benzocaine, a topical anesthetic found in some OTC products. The organization has determined the internally derived OEL to be 100 $\mu g/m^3$. The hygienist is granted enough funds to acquire three samples, and the results come back as 17, 12, and 22 $\mu g/m^3$, respectively.

TABLE 4.4

Summary Table of Hypothetical Sampling Data and their Analysis

Result (ug/m³)	Log-Transformed Data ($y = \ln(x)$)	$(\bar{y} - y)^2$
17	2.8332	0.000912
12	2.4849	0.101187
22	3.0910	0.082944
	$\Sigma y = 8.4091$	$\Sigma(\bar{y} - y)^2 = 0.18504$

The first step to analyzing the data requires the hygienist to convert the data (Table 4.4). This is done by taking the natural logarithm of the data. Once converted, the average of the log-transformed data points ($\bar{y}$) is obtained:

$$\bar{y} = \frac{\Sigma y}{n} = \frac{8.4091}{3} = 2.803 \tag{4.1}$$

Next, the standard deviation (sd) of the log-transformed data is tabulated:

$$sd = \sqrt{\frac{\Sigma(\bar{y} - y)^2}{n-1}} = \sqrt{\frac{0.18504}{2}} = 0.3042 \tag{4.2}$$

Finally, the 95th percentile of the exposure profile can be estimated using Equation 4.3. Note that in this instance, the value of Z is 1.645 and is used for all 95th percentile point estimate calculations:

$$95^{th} \text{ Percentile} = \exp\left(\bar{y} + \left[Z \times sd\right]\right) = \exp\left(2.803 + \left[1.645 \times 0.3042\right]\right) = 27.2 \text{ ug/m}^3 \tag{4.3}$$

In this example, the 95th percentile point estimate would be 27.2 µg/m³, which is below the assigned OEL of 100 µg/m³. Additionally, the 95th percentile falls into an AIHA Exposure Category 2, as it is less than 50% of the OEL. For many decision makers who are not familiar with statistics, this would be enough data to make a decision and state with some degree of confidence that exposures to benzocaine are controlled and no risk is present to the operators since the value is well below the OEL.

But the calculated point estimate is only for the acquired data and represents the exposure profile based solely upon that data. If an additional three samples were taken a month later, they would undoubtedly be different with different variations between them, giving rise to a much different 95th percentile, possibly in excess of the OEL. How then, do we say with any confidence that exposures are below a certain level? More specifically, as industrial hygienists, how can we be 95% sure that 95% of all exposures are below the OEL? This is where upper tolerance limits (UTLs) come into play.

UTLs take into account the uncertainty surrounding the acquired data and informs the hygienist with a certain amount of confidence where the true 95th percentile would fall on a distribution curve. UTLs are calculated using Equation 4.4. Note that the value of K in Equation 4.4 varies based on the level of confidence desired in the calculation (anywhere from 75% to 99%) as to where a certain percentile lies in a distribution curve (anywhere from 75th percentile to 99th percentile) and also depends on the number of samples acquired. These values of K can be obtained from tables of standard statistical factors for one-sided tolerance limits. In this instance, since we are interested in knowing the UTL$_{95\%,95\%}$ (where the 95th percentile falls with 95% certainty) and our number of samples was 3, our K value is 7.655:

$$\text{UTL}_{95\%, 95\%} = \exp\left(\bar{y} + \left[K \times sd\right]\right) = \exp\left(2.803 + \left[7.655 \times 0.3042\right]\right) = 169 \text{ µg/m}^3 \tag{4.4}$$

From these calculations, the hygienist would say with 95% confidence that the true 95th percentile of the exposure distribution is 169 µg/m³ and the exposure would be classified as an overexposure, yet not a single data point was over the exposure limit. Herein lies the problem with traditional statistics and small sample sizes: with small sample sizes, the level of uncertainty is extreme. Many seasoned industrial hygienists would view the UTL$_{95\%,95\%}$ of this data and disregard it, as the samples obtained did not even come close to approaching this level.

A common workaround for this dilemma is to tabulate a UTL with lower confidence, typically around 70% confidence. This calculation gives percentiles more in line with observed data. Unfortunately, the reduced confidence isn't necessarily aligned with the AIHA methodology of ensuring that virtually all employees are protected from overexposures. From a risk management perspective, this fails to meet the intended purpose of protecting workers with certainty. For years, those in general industry and the pharmaceutical sector were forced to use this approach and make compromises on their decision making.

One way to overcome this issue is to obtain more samples. As the number of samples increases, the uncertainty decreases. Let's assume that the hygienist was able to collect an additional three samples, with their concentrations reported as 23, 14, and 19 µg/m³ (Table 4.5).

$$\bar{y} = \frac{\Sigma y}{n} = \frac{17.1281}{6} = 2.8547 \tag{4.5}$$

$$sd = \sqrt{\frac{\Sigma(\bar{y} - y)^2}{n-1}} = \sqrt{\frac{0.32643}{5}} = 0.2555 \tag{4.6}$$

$$95^{th} \text{ Percentile} = \exp(\bar{y} + [Z \times sd]) = \exp(2.8547 + [1.645 \times 0.2555]) = 26.4 \text{ ug/m}^3 \tag{4.7}$$

$$UTL_{95\%,95\%} = \exp(\bar{y} + [K \times sd]) = \exp(2.8547 + [3.707 \times 0.2555]) = 44.8 \text{ µg/m}^3 \tag{4.8}$$

The addition of three more data points shows an improvement to the 95th percentile point estimate (Equation 4.7) but a notable reduction to the UTL$_{95\%,95\%}$ (Equation 4.8). Indeed, the 95th percentile decreases only slightly and is essentially within the same value. There is little difference in the 95th percentile whether there are three samples or six. However, the UTL$_{95\%,95\%}$ is very different. With the addition of three more data points, the UTL$_{95\%,95\%}$ fell from 169 to 44.8 µg/m³. The revised UTL$_{95\%,95\%}$ is still higher than all other acquired data points but is far more aligned with the actual data than before. Indeed, the value obtained from Equation 4.8 would result in the process being

---

TABLE 4.5

A Summary of Additional Hypothetical Sampling Data and Their Analyses

Result (ug/m³)	Log-Transformed Data ($y = \ln(x)$)	$(\bar{y} - y)^2$
17	2.8332	0.000462
12	2.4849	0.136752
22	3.0910	0.055838
23	3.1355	0.078849
14	2.6391	0.046483
19	2.9444	0.008046
	$\Sigma y = 17.1281$	$\Sigma(\bar{y} - y)^2 = 0.32643$

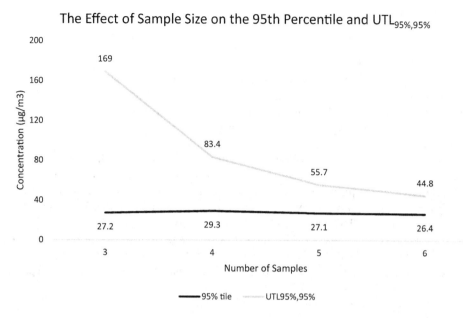

FIGURE 4.3   As more samples are acquired, the UTL95%, 95% exhibits a more drastic change than the 95th percentile.

classified as an AIHA Exposure Category 2, which result in the risk management team deeming the process acceptable and likely requiring no further actions.

As more samples are acquired, the estimated values of both the 95th percentile and the UTL$_{95\%,95\%}$ change, but the degree to which each changes is much more drastic. Figure 4.3 shows a comparison of these changes. The 95th percentile never changes by more than approximately 2.0 µg/m³ in value, indicating that the 95th percentile point estimate is relatively insensitive to additional data; that is, for consistent operations that do not drastically deviate, the 95th percentile remains fairly consistent even with the addition of more data points (this may not be the case for processes with significant variations in data, giving rise to large geometric standard deviations). But the 95th percentile, as mentioned previously, only examines the distribution based on the data at hand and not necessarily for all exposures.

In contrast, the change to the UTL$_{95\%,95\%}$ as more data are acquired is almost startling. With the addition of a single data point, the UTL$_{95\%,95\%}$ drops to 83.4 µg/m³, an approximately 50% decrease from the value with only three samples. Another significant drop occurs with the addition of a subsequent data points, with the value of the UTL$_{95\%,95\%}$ ultimately approaching but never reaching the tabulated 95th percentile. A few organizations and industries which utilize traditional statistics have realized this important trend and require a minimum of six samples to be acquired for SEG analysis.

Yet the drawback to such a mandate is obvious in that the sampling budget has now doubled in terms of cost and time. For smaller organizations or those whose IH departments are still up-and-coming, the physical budget for such analyses is simply not there. It is a common occurrence for the bare minimum number of samples to be acquired and to compile statistics to make decisions. A better approach would be a methodology that combines statistical applications with professional judgment of industrial hygienists while sufficiently providing data to make risk-based decisions. This is where Bayesian statistics offers a practical solution.

The data presented for the preceding example is that for a very well-controlled process wherein the data points are fairly close together; that is, the results are very consistent with each other. Reality often presents a much different result, with a wide variation in the collected data. In such instances, the same overall trend occurs: the UTL$_{95\%,95\%}$ drops considerably with more data points and "approaches" the tabulated 95th percentile of the data, but the gap between these values can be considerably larger than what is presented in our example. This is often more representative of the real world.

4.3.3.2   Bayesian Statistics

Far and away, the most common means of classifying occupational exposures within various pharmaceutical and consumer healthcare organizations is the use of Bayesian statistics. A full discussion on the theory and development of Bayesian statistics is beyond the scope of this book, but a practical description and treatment of the process will be described.

The fundamental aspect of Bayesian statistics is that the method can be used to update our previous beliefs with additional information; that is, if we have a previously believed idea (or data set), we can further refine it when new data is added to the mix. The theorem is broken down into three parts referred to as the prior, the likelihood, and the posterior (see Equation 4.9). The prior is defined as $P(A)$ and is essentially the believed probability of event A occurring. The term $P(B|A)$ is referred to as the likelihood. This term considers the probability of an event B occurring given that probability of event A is true. This is a critical concept since it depends on event B occurring only if event A occurs. The term $P(A|B)$ in Equation 4.9 is the posterior, and this represents the updated beliefs based on the new evidence. The denominator $P(B)$ is a marginalization factor and represents the probability of event B being true.

The fundamentals of Bayes' theorem.

$$P(A|B) = \frac{P(B|A) \times P(A)}{P(B)} \tag{4.9}$$

Bayesian statistics is a method to calculate probabilities based off Equation 4.9. The use of Bayesian statistics has found applications in a variety of industries and settings, allowing users to solve problems that were once difficult or impossible. One of the earliest implementations of Bayes' theorem into industrial hygiene was in 2006 by Hewett and colleagues.[9] They described a sophisticated means of analyzing the probability of where the 95th percentile of an exposure profile would reside which aligned with the AIHA exposure classification scheme. By utilizing Bayes' theorem and adapting to an industrial hygiene setting, it circumvented much of the guesswork surrounding uncertainty and sample sizes.

The utility of Bayesian statistics can be demonstrated by application to the data shown in Table 4.4 (Figure 4.4). For most hygienists assessing a potential exposure during a given process for the first time, they do not have any data to say how the exposure profile will look. That is, there is no data to influence their initial beliefs. From a Bayesian perspective, these initial beliefs constitute the "prior" function of Bayes' theorem. When there is no available data, the hygienist typically assumes a "uniform prior"; that is, there is an equal probability of the 95th percentile residing in each of the five AIHA exposure categories. This is represented in Figure 4.4 by having all probabilities in the Prior field as 0.2. Once our data are entered, the Likelihood is calculated. The likelihood term is a factor of the data in relation to the previously believed data; that is, the output of the Likelihood

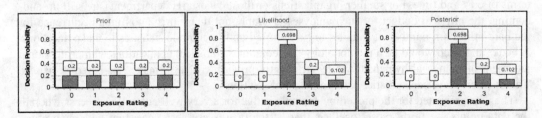

FIGURE 4.4   A graphical representation of the prior, likelihood, and posterior probability charts in Bayesian statistics for data shown in Table 4.4. Note that the likelihood and posterior are identical because a uniform prior was utilized.

is partially dictated by the previous probabilities. The final output is the Posterior, which is the updated probabilities of occurrence. If a uniform prior is used, such as in our example, the Posterior and Likelihood will have the same output values.

In our example, the likelihood rating gives an output of 0.698 in Exposure Category 2, with significantly lower probabilities for Categories 3 and 4. Utilizing this methodology, the practicing hygienist is likely to conclude that the true 95th percentile for this exposure profile is a Category 2. The categorization allows the hygienist to prioritize this exposure with more confidence than using classical statistics. Furthermore, knowing that a Category 2 exposure is one wherein the 95th percentile approaches 50% of the OEL (which is 50 µg/m³ in our example) is much more closely aligned with the observed data. Of critical importance is the fact that same risk categorization (Category 2) is obtained from Bayesian analysis using three data points as with traditional statistics using six data points. If the same conclusions can be reached, it is generally preferable to use the method that requires half the resources.

As stated above, our example utilized a uniform prior. This is a reasonable approach when no previous data is available regarding the exposure in question. However, once sampling and analysis are performed, the output can then be used as a prior for future assessments. In this manner, routine follow ups and assessments continually refine the exposure profile over time. This constant reassessment over time fits well into the ISO 31000 methodology of continually managing risk. Describing this particular aspect of Bayesian statistics as an advantage would be an understatement. When traditional statistics are utilized, every time the process is essentially evaluated anew, never taking prior data into account. Each analysis is unique and independent, and does nothing to refine the characterization of the exposure profile to the SEG. Bayesian statistics circumvents this by bringing in past relevant data to continually refine and improve the characterization of the risk surrounding the process and SEG.

The practicality and ease of use has made Bayesian statistics the preferred method of assessing industrial hygiene data within the pharmaceutical and consumer healthcare industries. The advantages are significant. It immediately adopts the AIHA exposure categories protocol, is easy to implement, gives more realistic data outputs over traditional statistics, and permits continual refinement and risk management.

Another appeal to the use of Bayesian statistics is the application of industrial hygiene professional judgment when developing a prior probability. Even though we mentioned in Chapter 3 that when hygienists attempt to estimate an exposure profile, they are often incorrect and often underestimate the true value, it is nevertheless a common occurrence for practicing hygienists to still put their "expert judgment" on an exposure. Many will attempt to gauge an exposure profile by simply observing how a process is performed and utilizing prior knowledge and experience to ascertain an exposure level, and this is a perfectly acceptable practice. However, it is up to each organization to define upfront how a prior will be utilized during a risk assessment. It is not uncommon for new processes or baseline evaluations to utilize a uniform prior and then to use the subsequent output as the new prior during subsequent reanalysis. Regardless of the tactic to be employed, the organization and risk assessors should agree on the standard approach prior to beginning or analyzing any data.

It almost goes without saying that Bayesian statistics provides a realistic, almost traditional risk management treatment of industrial hygiene processes. The use of probabilities as to where the 95th percentile resides is a quantum leap in risk characterization. Rather than relying on subjective assessments of a process based on vague definitions designed to be "one size fits all", Bayesian statistics uses representative data (acquired via means outlined in Chapter 3) to be compared against data-driven toxicologically derived occupational exposure limits (as described in Chapter 2) to derive probabilities of where the 95th percentile is likely to be relative to the pre-defined AIHA exposure categories. This is an extremely powerful tool for risk managers, as it can directly relate to a given risk appetite as well as clearly dictate courses of action based on the findings.

A final benefit to the use of Bayesian analysis is the capability to analyze censored data sets. In Chapter 3, we discussed how "non-detects" are represented as concentrations that are "less than (<)" the reporting limit. Such instances occur even with the most sensitive of analytical methods. When censored data sets occur, traditional statistics cannot be performed. To circumvent this, many hygienists have employed seemingly arbitrary tools to arrive at usable numbers. These can include dividing the censored result by 2 or dividing the result by $\sqrt{2}$. However, these methods can lead to significantly biased and skewed results. In contrast, Bayesian analysis utilizes a methodology called maximum likelihood estimation (MLE), which has been shown to give more accurate results compared to other protocols with handling censored data sets.[10] The use of Bayesian analysis has greatly facilitated the risk characterization of processes which routinely provide censored data sets, such as during surrogate testing of containment enclosures and other analyses which have insensitive analytical methods. Perhaps this single point of being able to perform analyses with censored data points makes the Bayesian approach more desirable than other approaches and has made it the preferred data analysis tool in the pharmaceutical industry.

4.4 SUMMARY

Industrial hygiene risk assessments have come a long way. Historically such assessments were largely qualitative in nature and utilized subjective criteria for both hazard evaluations and exposure evaluations. Humans are, by and large, not very good at estimating exposures, and this becomes even more difficult when observing tasks involving extremely small quantities of material. Risk matrices are often used to evaluate risks in a qualitative or semi-quantitative manner, but these often provide inaccurate results due to a number of factors. It is also difficult to apply a risk matrix to industrial hygiene in an accurate manner as well. To overcome such inconsistencies and to inject objectivity, the pharmaceutical industry has adopted a data-driven approach to assessing risk. The modern risk assessment paradigm keeps the spirit of the original NRC human health risk assessment by evaluating both hazards and exposures to an exposure group. As many substances in the industry do not have OELs assigned to them, the universally accepted practice of exposure banding enables risk assessors to gauge an acceptable airborne level for substances. The adoption of Bayesian statistics has enabled practitioners to be able to more confidently assign exposure profiles to small data sets, thereby largely bypassing the need for larger sampling budgets while still obtaining useful data. Hygienists still often grapple with the use of risk matrices as these are often adopted by organizations to numerically and visually categorize risks of all kinds, but the conversion of industrial hygiene data into arbitrarily defined ordinal data groups often presents technical as well as logical challenges in addressing risks. The acquisition of hazard groupings and exposure assessments, two pieces of the risk puzzle, organically creates a straightforward means of assigning exposure categories for an SEG, thereby driving the risk assessment and subsequently the risk management process. The modern practice aligns with both the NRC recommendations and the ISO 31000 framework that is being adopted by organizations around the world. Regardless of the analytical method chosen, once a risk level has been assigned to the SEG for the process or task, the next step is to decide on how best to treat the risk. We will delve deeper into industrial hygiene risk treatment in subsequent chapters.

NOTES

1 NRC. *Risk Assessment in the Federal Government: Managing the Process.* Washington, DC: National Academy Press; 1983.
2 NRC. *Understanding Risk: Informing Decisions in a Democratic Society.* Washington, DC: National Academy Press; 1996.

3 Science Policy Council. (2000, December). *Risk Characterization Handbook*. (EPA 100-B-00-002). Environmental Protection Agency.

4 Leidel, N., Busch, K., and Lynch, J. (1977). *NIOSH Occupational Exposure Sampling Strategy Manual*. Department of Health and Human Services.

5 Hubbard, D. W. (2014). *How to Measure Anything: Finding the Value of Intangibles in Business*. John Wiley & Sons.

6 Hubbard, D. W. and Evans, D. (2010). Problems with scoring methods and ordinal scales in risk assessment. *IBM Journal of Research and Development*, *54*(3), 2:1–2:10.

7 Hubbard, D. W. (2020). *The Failure of Risk Management: Why It's Broken and How to Fix It*. John Wiley & Sons.

8 Cox Jr., L. A. (2008). What's wrong with risk matrices? *Risk Analysis: An International Journal*, *28*(2), 497–512.

9 Hewett, P., Logan, P., Mulhausen, J., Ramachandran, G., and Banerjee, S. (2006). Rating exposure control using Bayesian decision analysis. *Journal of Occupational and Environmental Hygiene*, *3*(10), 568–581.

10 Hewett, P. ans Ganser, G. H. (2007). A comparison of several methods for analyzing censored data. *Annals of Occupational Hygiene*, *51*(7), 611–632.

5 Risk Treatment
Control Banding

5.1 INTRODUCTION

The first several chapters of this book dealt with laying out the necessary criteria to assess worker risks of occupational exposure to chemicals and APIs as accurately as possible, particularly with the constraints that are typically placed upon the practicing industrial hygienist. By identifying the particular hazards of a substance or placing it into an occupational exposure band if need be allows the hygienist to conduct the first part of a human health risk assessment. The proper outlining of a sampling plan then allows for the acquisition of data to evaluate the exposure assessment for the individual and SEG to ascertain the overall risk of exceeding the designated threshold limits. In all this practice represents the first portion of the risk management methodology, the risk assessment.

The next step in the ISO 31000 risk management methodology is risk treatment (Figure 5.1). Conceptually, this step requires risk managers to evaluate all options for dealing with an identified risk. Among risk managers, these options generally include:

- **Avoidance** – this strategy is one adopted by risk managers when they don't wish to take on the risk to their company or organization. Avoidance occurs when the activity that gives rise to the risk is not undertaken;

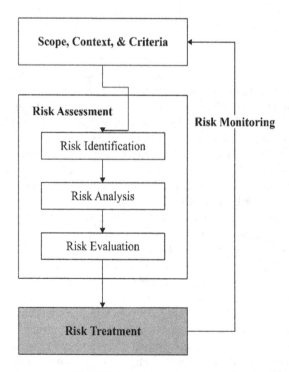

ISO 31000 Methodology

FIGURE 5.1 One of the final steps of the ISO 31000 risk management flow is risk treatment.

DOI: 10.1201/9781003273455-5

- **Transfer** – the transfer strategy for risk treatment occurs when the activity is not avoided, but the risk no longer directly affects the parent organization;
- **Acceptance** – perhaps the simplest of all the main strategies presented, acceptance of a risk occurs when the assessing organization weighs all the alternatives and decides that risk itself, should it come to fruition, can be absorbed by the company or has a minimal impact to its mission, vision, or bottom line. This route essentially requires the organization to do nothing;
- **Reduction** – if acceptance is the simplest of risk treatment options, then reduction is the most difficult of all the strategies presented. Reduction of a risk requires additional planning and actions to be undertaken in order to reduce either the severity of a hazard or, more likely, reduce the likelihood of a hazard occurring.

The four risk treatment strategies presented above are high-level concepts which can be applied to any risk within any industry, including the pharmaceutical and consumer healthcare industries. But it is important to remember that from an industrial hygiene perspective, we are dealing with human health risk assessments, and therefore, some of the risk treatment options are not generally viable. Avoidance is not typically an available option, as the product being made and marketed is what gives rise to the risk in the first place. To implement avoidance would be to ask the organization to not make a particular drug or product and given the monetary implications of not making a certain product (especially if it is expected to bring a significant return), avoidance is almost guaranteed to not be a viable risk treatment option.

Risk transfer may be a suitable risk treatment option within the pharmaceutical industry. Traditionally risk transfer involves passing the risk on to a designated third party, and in most industries, this often takes the form of insurance. However, another form of risk transfer exists that is uniquely applicable to the pharmaceutical and consumer healthcare sectors. Outsourcing, or hiring a third party to perform the high-risk activity on your behalf, is a commonly utilized route to transfer risk. Situations in which outsourcing presents a viable strategy include temporarily increasing production capacity or, in the context of industrial hygiene, when a product needs to be produced and the necessary safeguards or containment levels to handle toxic or potent substrates cannot be achieved with the equipment on hand.

In contrast to other risk treatment options, risk reduction requires a significant amount of time and capital investment in addition to technical expertise. Reducing risk in the context of industrial hygiene involves implementing the hierarchy of controls (Figure 5.2). The hierarchy of controls is a well-established paradigm in the safety field by which hazards can be mitigated by implementation of various controls. The more effective controls, elimination and substitution, are located at the top of the pyramid and are more difficult to implement or less commonly utilized. We have already touched on how elimination and substitution are not typically viable risk treatment options, so we will not focus on these controls in the hierarchy. Unquestionably, the most commonly employed controls within the pharmaceutical and consumer healthcare industries are engineering controls, administrative controls, and utilization of PPE.

But once a risk is adequately identified and characterized, which controls should be used? There are a myriad of options within the three primary categories of controls from which to choose, and the list only seems to grow. Historically, managers would simply put operators into additional layers of PPE. This is no longer an accepted practice since PPE only function if the operators utilize them correctly, plus there are additional recurring costs associated with PPE use. The modern approach to protecting employees by reducing exposure risks is to utilize engineering controls. These are devices or equipment which either removes the hazard(s) from the working environment before they can do harm to employees or they completely prevent their release. On the other hand, it is entirely possible to over-engineer a process and install unnecessary equipment and safeguards. Furthermore, engineering controls require significant capital expenditure and time to install, a luxury that cannot always be afforded in some organizations.

Hierarchy Of Hazard Controls

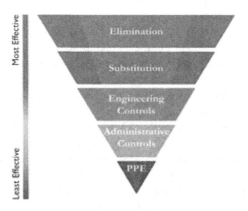

FIGURE 5.2 The NIOSH hierarchy of controls stresses elimination and substitution as primary means of protection, but engineering and administrative controls are more pragmatic, while PPE is considered a last line of defense.

Knowing which controls to utilize based on the tasks performed and the nature of materials being handled is an exceptional advantage for the risk management team.

The practicing hygienist is expected to be proficient in not only selecting the appropriate controls for a given operation and risk but assisting with its implementation and operator training as well. This chapter will focus on the tools and methods typically used within the pharmaceutical and consumer healthcare industries to select the appropriate level of controls for reducing exposure risks to employees. When used appropriately, these tools help to meet the anticipated goal of risk treatment by minimizing employee exposures to airborne contaminants.

5.2 BEFORE SELECTING AN ENGINEERING CONTROL

Once the decision has been made by the risk management team that a risk needs to be effectively treated (e.g., reduced), a decision must then be made as to *how* the risk will be reduced. Not all engineering controls are equal: some are far more efficient at removing contaminants than others, whereas some are suitable for particular processes and not others. The sheer number of engineering controls on the market is nothing short of staggering. The modern practice of control banding within the pharmaceutical industry greatly facilitates the selection of engineering controls for a given class of substances. Furthermore, the nature of the materials being handled/controlled also play a key role. In this manner, a thorough understanding of the material including its associated hazards as well as intrinsic and extrinsic properties will assist the design team in selecting an appropriate piece of equipment. Finally, when any sort of change is introduced to an existing process (e.g., retrofitting), there will undoubtedly be downstream effects on the process itself. Collectively understanding these components will greatly assist in choosing the best engineering control for a process.

5.2.1 PHARMACEUTICAL CONTROL BANDING

5.2.1.1 Performance-Based Exposure Control Limits

A recurring theme within this book has been the notion that APIs have become more potent over time; that is, patients who require a prescription drug for a particular ailment need to take less of the drug to elicit the desired therapeutic effect. While this is a boon for patients, it presents a hazard to the pharmaceutical operators who handle and package the medications in their raw (unfinished) form. And as the potencies have increased, the requisite OELs to which operators can be exposed have decreased in kind.

Pharmaceutical companies readily recognized the need to protect their workers while handling such increasingly potent substances, and as a result, the industry developed the idea of control banding.[1] The concept of control banding is essentially the same as exposure banding (see Chapter 2), but with the added extension that a series of high-level recommended controls are added to each band. The concept was pioneered in the 1990s and found to have tremendous success and has since been adopted as a general tool that can be applied to essentially any industry.

One of the first published descriptions of the control banding process was put forth by Merck & Co. (MSD).[2] Referred to at the time as performance-based exposure control limits (PB-ECL), the authors described not only their classification scheme for banding substances based on available toxicological information, but they implemented controls for achieving the desired exposure range (see Table 5.1). The requirements put forth by MSD for achieving specific levels of operator exposure was a massive leap forward for those in similar industries and working with similar materials. As would be expected, the number of controls increases substantially as the exposure category increases and the allowable exposure levels decrease. For instance, chemical protective clothing is only recommended for PB-ECL 3 but is required for PB-ECL 4, and double layering of such clothing is required for PB-ECL 5. Importantly, the criteria for the PB-ECL mantra were based off years of industrial hygiene sampling data and essentially trial-and-error with various levels of controls to achieve the desired targets. Additionally, all exposure categories required some sort of industrial hygiene monitoring to characterize the exposure (this concept will come up again later). The result was a first of its kind blueprint off which anyone in the pharmaceutical industry can build.

5.2.1.2 Modern Pharmaceutical Control Banding

In 2002, authors Nigel Hirst, Mike Brocklebank, and Martyn Ryder published *Containment Systems: A Design Guide*, a standalone book detailing pharmaceutical industry-specific control banding method.[3] The authors largely based their system on the foundation of COSHH Essentials, a control banding system developed for small and medium-sized enterprises within the UK. The authors developed the system in conjunction with Steve Maidment, one of the primary architects of the COSHH Essentials system. Yet there are distinct differences worth noting.

The first difference is regarding the exposure banding paradigm. There were six control bands, not five as found within COSHH, classified as A through F. Each band had an allowable exposure range to which it was believed that if airborne exposures were met, then workers would be protected from the associated hazards of that substance. Control bands A, B, and C aligned with COSHH Essentials regarding the exposure limits, but bands D, E, and F were distinctly new in that they provided substantially lower exposure ranges. This was due to the realization that increasingly potent compounds required distinct categories and, more importantly, different forms of risk treatment. Moreover, some of the R-phrases were redistributed among the control bands owing to the newly expanded exposure limits. As mentioned in the previous section, the R-phrases are now obsolete and have since been replaced by the GHS H-phrases. Table 5.2 shows the exposure banding concept along with the associated risk phrases and how they relate to each other. The H-phrases presented in Table 5.2 were not part of the original publication but are presented here to provide the reader with contemporary context.

An examination of the anticipated exposure limits in control bands E and F are reflective of the need within the pharmaceutical industry for its own control banding scheme. COSHH Essentials only included exposure limits as low as $10 \ \mu g/m^3$ and required specialist advice for anything lower; in contrast, the pharmaceutical industry frequently deals with potent compounds and highly potent compounds possessing anticipated allowable exposure ranges that are several orders of magnitude lower than those encountered within COSHH Essentials.

But there were additional considerations placed into the revised control banding system. Significantly, the aspects relating to volume/quantity of material handled, the duration of the task,

TABLE 5.1

The PB-ECL System as Originally Detailed by MSD[2]

Design Consideration	PB-ECL Category (Potency Level)				
	1 (>100 mg/day)	2 (>10–100 mg/day)	3 (0.1–10 mg/day)	4 (<0.1 mg/day)	5 (<0.1 mg/day)
General concept	No unauthorized entry; all work surfaces cleaned at end of each day; eating, drinking, smoking, and cosmetic application prohibited; use of lab coats and work uniforms is recommended	No unauthorized entry; all work surfaces cleaned at end of each day; eating, drinking, smoking, and cosmetic application prohibited; use of lab coats and work uniforms is strongly recommended	Controlled access to the work area is strongly recommended; allowed personnel should undergo specific training to be granted access; all work surfaces must be decontaminated after all high-risk activities; eating, drinking, smoking, and cosmetic application prohibited; strict adherence to work practices must be implemented and enforced without tolerance for deviation; signs must be posted relating the hazards of the materials	Controlled access to the work area is required; allowed personnel should undergo specific training to be granted access; all work surfaces must be decontaminated after all high-risk activities; eating, drinking, smoking, and cosmetic application prohibited; strict adherence to work practices must be implemented and enforced without tolerance for deviation; signs must be posted relating the hazards of the materials	Controlled access to the work area is required; allowed personnel should undergo specific training to be granted access; all work surfaces must be decontaminated after all high-risk activities; eating, drinking, smoking, and cosmetic application prohibited; strict adherence to work practices must be implemented and enforced without tolerance for deviation; signs must be posted relating the hazards of the materials
Containment level	No special containment equipment is required; LEV should be utilized	No special containment equipment is required; LEV should be utilized	Open handling must be limited to very small quantities; fume hoods and open-face containment equipment are suitable with average face velocities of 80 ft/min; use of other containment technologies is recommended to prevent contamination into uncontrolled areas	Open handling prohibited; containment technology is required for all operations; gloveboxes, isolators, and transport systems required	Open handling prohibited; containment technology is required for all operations; gloveboxes, isolators, and transport systems required; all equipment must be leak tested to show 100% leak proof; robotic intervention recommended
General ventilation	≥7 air changes per hour; recirculation of air is permitted in production areas if adequate filtration is present,	≥7 air changes per hour; recirculation of air is permitted in production areas if adequate filtration is present,	≥10 air changes per hour; recirculation of air is permitted in only limited scenarios; air flow must be away from employee breathing zones; work space must be under negative pressure relative to adjacent areas	≥12 air changes per hour; recirculation of air is strictly prohibited; air flow must be away from employee breathing zones; workspace must be under negative pressure relative to adjacent areas;	≥12 air changes per hour; recirculation of air is strictly prohibited; air flow must be away from employee breathing zones; workspace must be under negative pressure relative to adjacent areas;

(Continued)

TABLE 5.1 (Continued)
The PB-ECL System as Originally Detailed by MSD[2]

Design Consideration	PB-ECL Category (Potency Level)				
	1 (>100 mg/day)	2 (>10–100 mg/day)	3 (0.1–10 mg/day)	4 (<0.1 mg/day)	5 (<0.1 mg/day)
	and exposures are <50% of the OEL (AIHA Category 2 or lower); no recirculation in non-production areas	and exposures are < 50% of the OEL (AIHA Category 2 or lower); no recirculation in non-production areas		air locks with interlocked doors are required to access containment area	air locks with interlocked doors are required to access containment area; double air locks recommended
Local exhaust ventilation (LEV)	Must be present at the source and follow ACGIH design criteria; recirculation is permitted through HEPA filters only	Must be present at the source and follow ACGIH design criteria; recirculation is not permitted	Must be present at the source and follow ACGIH design criteria; exhaust must be HEPA filtered out the building	Must be present at the source and follow ACGIH design criteria; exhaust must be HEPA filtered out the building	Must be present at the source and follow ACGIH design criteria; exhaust must be HEPA filtered out the building
Surfaces	No special requirements	Easily cleaned surfaces is required	Smooth and non-porous surfaces with minimal ledges; all surfaces must be easily cleaned	Smooth and non-porous surfaces with minimal ledges; all surfaces must be easily cleaned; all surfaces must be contiguous	Smooth and non-porous surfaces with minimal ledges; all surfaces must be easily cleaned; all surfaces must be contiguous
Maintenance, cleaning, waste disposal, and decontamination	Floor sweeping is prohibited; all powder spills must be cleaned using vacuum; decontamination is not required for validated surface testing; waste is	Floor sweeping is prohibited; all powder spills must be cleaned using vacuum; decontamination is not required for validated surface testing; waste is	Floor sweeping is prohibited; cleaning is conducted with HEPA vacuums prior to wet cleaning; decontamination is not required for validated surface testing; waste is double-bagged; no crushing or shredding of waste is allowed; clean-in-place (CIP) systems are recommended for equipment; safe	Floor sweeping is prohibited; cleaning is conducted with HEPA vacuums prior to wet cleaning; decontamination is recommended for surfaces; surface testing is required for delisting equipment or removing it from the area; waste is double-bagged; no crushing or shredding of waste is allowed;	Floor sweeping is prohibited; cleaning is conducted with HEPA vacuums prior to wet cleaning; decontamination is recommended for surfaces; surface testing is required for delisting equipment or removing it from the area; waste is double-bagged; no crushing or shredding of waste is

(Continued)

TABLE 5.1 (*Continued*)
The PB-ECL System as Originally Detailed by MSD[2]

Design Consideration	PB-ECL Category (Potency Level)				
	1 (>100 mg/day)	2 (>10–100 mg/day)	3 (0.1–10 mg/day)	4 (<0.1 mg/day)	5 (<0.1 mg/day)
	double-bagged; no crushing or shredding of waste containers is allowed	double-bagged; no crushing or shredding of waste containers is allowed	change filters are required on ventilation systems; maintenance access should be provided outside the containment area	clean-in-place (CIP) systems are required for equipment; safe change filters are required on ventilation systems; maintenance access should be provided outside the containment area	allowed; clean-in-place (CIP) systems are required for equipment; safe change filters are required on ventilation systems; maintenance access should be provided outside the containment area
PPE	Respiratory protection commensurate with exposure levels is required; minimum of qualitative fit testing is required for tight-fitting respirators; effective gloves must always be worn; appropriate eye protection must be worn	Respiratory protection commensurate with exposure levels is required; minimum of qualitative fit testing is required for tight-fitting respirators; effective gloves must always be worn; appropriate eye protection must be worn	Only powered air purifying respirators (PAPRs) or supplied air hoods are allowed for respiratory protection; for PAPRs, HEPA filters must be used; effective gloves must always be worn; appropriate eye protection must be worn; chemical protective outer garments suitable for material being handled are required	Only powered air purifying respirators (PAPRs) or supplied air hoods are allowed for respiratory protection; for PAPRs, HEPA filters must be used; effective gloves must always be worn; appropriate eye protection must be worn; chemical protective outer garments suitable for material being handled are required; double outer protective chemical garment is recommended	Only powered air purifying respirators (PAPRs) or supplied air hoods are allowed for respiratory protection; for PAPRs, HEPA filters must be used; effective gloves must always be worn; appropriate eye protection must be worn; chemical protective outer garments suitable for material being handled are required; double outer protective chemical garment is required
IH monitoring	Full-shift breathing zone samples are required to accurately characterize the workplace; routine monitoring is not	Full-shift breathing zone samples are required to accurately characterize the workplace; the	Full-shift breathing zone samples are required to accurately characterize the workplace; the sampling plan should utilize a statistical valid number of samples; routine monitoring is required to ensure all	Full-shift breathing zone samples are required to accurately characterize the workplace; the sampling plan should utilize a statistical valid number of samples; routine monitoring is	Full-shift breathing zone samples are required to accurately characterize the workplace; the sampling plan should utilize a statistical valid number of samples; routine monitoring is

(*Continued*)

TABLE 5.1 (*Continued*)
The PB-ECL System as Originally Detailed by MSD[2]

Design Consideration	PB-ECL Category (Potency Level)				
	1 (>100 mg/day)	2 (>10–100 mg/day)	3 (0.1–10 mg/day)	4 (<0.1 mg/day)	5 (<0.1 mg/day)
	required unless the action limit is exceeded	sampling plan should utilize a statistical valid number of samples; routine monitoring is required to ensure all controls and work practices are effective; ventilation testing is required	controls and work practices are effective; ventilation testing is required	required to ensure all controls and work practices are effective; ventilation testing is required	required to ensure all controls and work practices are effective; ventilation testing is required
Medical surveillance	General chemical/ pharmaceutical operator surveillance should be performed as determined by a health professional	Surveillance should be based on anticipated health effects of the material being handled, potential for exposure, and availability of test criteria	Surveillance should be based on anticipated health effects of the material being handled, potential for exposure, and availability of test criteria	Surveillance should be based on anticipated health effects of the material being handled, potential for exposure, and availability of test criteria	Surveillance should be based on anticipated health effects of the material being handled, potential for exposure, and availability of test criteria

TABLE 5.2

Exposure Banding Categorization as Proposed by Hirst, Brocklebank, and Ryder[7]

Control Band	Exposure Limit	R-Phrases	GHS H-Phrases
A	1,000–10,000 µg/m³ (dust) 50–500 ppm (vapor)	R36, R38	H319, H315
B	100–1,000 µg/m³ (dust) 5–50 ppm (vapor)	R20, R21, R22 (except with R48)	H332, H312, H302
C	10–100 µg/m³ (dust) 0.5–5 ppm (vapor)	R23, R24, R25 (except with R48) R34, R35, R37, R41, R43, R48/20, R48/21, R48/22	H331, H311, H301, H314, H335, H318, H317, H373
D	1.0–10.0 µg/m³ (dust) 0.05–0.5 ppm (vapor)	R26, R27, R28 Carc/Mut. Cat. 3 R40 R48/23, R48/24, R48/25 R60, R61, R62, R63	H330, H310, H300 H351 H372 H360
E	0.01–1.0 µg/m³ (dust) 0.005–0.05 ppm (vapor)	R42, R45, R46, R49	H334, H350, H340, H350
F	<0.01 µg/m³ (dust) <0.005 ppm (vapor)	None assigned	None assigned

and characterization of powder dustiness or liquid volatility were all taken into consideration. From a risk perspective, the banding and categorization aspect applied to the "hazard" variable of risk, but the inclusion of the additional content addressed the "exposure" variable. Regarding volume or quantity of material handled, the authors considered three general scales of chemical use: lab scale/pilot plant, mid-range production, and large-scale production, classified as small (gram scale), medium (kilogram scale), and large (ton scale). Just as COSHH Essentials logically assumed, if an employee physically handles larger quantities of a substance, there is greater propensity or probability of exposure to that substance.

The next significant leap was the consideration of the likelihood of the material to become airborne. After all, if a substance is not able to become airborne, then there is virtually no inhalation exposure risk. Solid materials that exist primarily as pellets, pills, dense chunks, or other non-friable forms were deemed "low" dustiness, whereas typical crystalline materials were classified as "medium" dustiness. This was due to the fact that upon handling, dust can often be observed during handling of crystalline substances (think pouring table sugar), but these often settle out of the air relatively quickly. Finally, very fine and light powders were classified as "high" dustiness since once they are airborne, the resulting particulates can remain airborne for a significant amount of time, increasing the likelihood of exposure.

Hirst, Brocklebank, and Ryder also took the duration of a task or process into account. This was a fundamental difference from COSHH Essentials which originally did not factor process duration (this is no longer the case, as the online user tool specifically requires a task duration component). Again, the logic behind the inclusion made intuitive sense: the longer a person performs a task, the more chance that exists for an employee to become exposed. For the purposes of the assessment, the authors estimated that a "short" task was one which took less than 30 minutes, and a "long" task was anything which exceeded the 30-minute timeframe. It was not clear if the "long" label also applied to cumulative time, such as several 5- or 10-minute processes being performed repeatedly throughout the day.

When factored together, all three variables were assembled into a matrix (Table 5.3). The resulting output were various "exposure potentials", or EP levels. The EP levels were assigned values, 1 through 4. The higher the EP level, the higher the "risk" for exposure during said process. This qualitative (and subjective) assessment of exposure allowed risk managers to quickly ascertain what

TABLE 5.3

Exposure Potential for Dusts and Powders

Quantity	Dustiness Potential					
	Low		**Medium**		**High**	
Small (grams)	EP1	EP1	EP1	EP2	EP2	EP3
Medium (kg)	EP1	EP2	EP2	EP3	EP3	EP4
Large (tons)	EP2	EP3	EP3	EP4	EP3	EP4
Duration	Short	Long				

TABLE 5.4

Required Control Strategy for Handling Dusts and Powders

Hazard Band	Exposure Potential (EP)			
	EP1	**EP2**	**EP3**	**EP4**
A	Strategy 1	Strategy 1	Strategy 1	Strategy 2
B	Strategy 1	Strategy 2	Strategy 2	Strategy 3
C	Strategy 2	Strategy 3	Strategy 3	Strategy 4
D	Strategy 3	Strategy 3	Strategy 4	Strategy 4
E	Strategy 4	Strategy 4	Strategy 4	Strategy 4
F	Strategy 5	Strategy 5	Strategy 5	Strategy 5

level of "risk" existed for a given process based solely on the nature of the materials being handled, their properties, and the duration of the process.

The exposure potential factor serves as the "probability" or "likelihood" factor when assessing risk and is used in the final step of the modified control banding methodology. When paired with the identified hazard band of the substance being handled, another matrix is obtained (Table 5.4). The outcome of this final matrix is the resultant control strategy to be used for that process. In total, there are five control strategies, listed numerically as 1 through 5, and each corresponds to increasing level of control.

A similar strategy is employed when liquids are utilized. The first factor to assess is how well the material can become airborne. Unlike solid materials, liquids can be classified by their relative volatility. This can be accomplished by either comparing vapor pressures or boiling points, but vapor pressures are often harder to come by than boiling points. Hirst, Brocklebank, and Ryder utilized the same volatility assessment methodology as COSHH Essentials, which classifies liquids according to their boiling points (at one atmosphere of pressure) and the temperature at which the operation will be performed. From this classification, liquid volatility can be classified as "low", "medium", or "high" (Figure 5.3).

Just as the solid handling scheme required the user to identify the scale of use, so too does the liquid handling scheme. The scale for liquids is broken into three categories: small (mL) quantities, medium (liters) quantities, and large (cubic meters) quantities. When paired with the volatility of the liquid, yet another matrix is derived, the outcome of which is the "exposure potential" for liquid handling (Table 5.5). Just as the solid exposure potential matrix (Table 5.3) utilized a time component, so too does the liquid exposure potential matrix. This aids in delineating between control

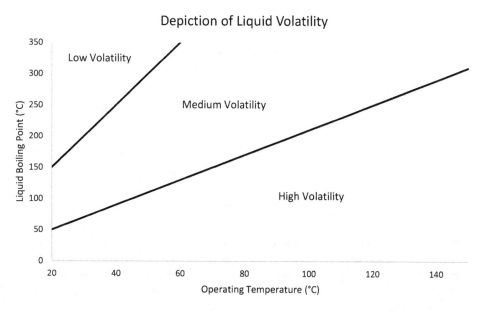

FIGURE 5.3 A graphical depiction of the liquid volatility classification used in COSHH Essentials and the modified control banding scheme.

TABLE 5.5
Exposure Potential Matrix for Liquids

Quantity	Liquid Volatility					
	Low		Medium		High	
Small (mL)	EP1		EP1		EP2	
		EP1		EP2		EP3
Medium (liters)	EP1		EP2		EP3	
		EP2		EP3		EP3-4
Large (cubic meters)	EP2		EP3		EP3	
		EP3		EP4		EP4

---

strategies based on duration. Once the exposure potential for liquids has been derived, it is then paired with the hazard band in a final matrix to determine the requisite control strategy (Table 5.6).

The output of the final control band matrices are various recommended control strategies:

- **Control Strategy 1** – Controlled general ventilation
- **Control Strategy 2** – Local exhaust ventilation (LEV)
- **Control Strategy 3** – Open handling within an isolator
- **Control Strategy 4** – Closed handling within an isolator
- **Control Strategy 5** – Remote robotic handling

The control strategies are somewhat conceptually similar to the output of COSHH Essentials. However, the control strategies put forth by the authors largely depart from the general, high-level recommendations and instead return to the more granular type of recommendations put forth in the PB-ECL methodology from MSD. For instance, each control strategy lists specific classes of engineering controls that are appropriate for controlling exposures to the desired airborne concentrations of the particular band of interest. Furthermore, additional guidance on maintenance, PPE

TABLE 5.6
Required Control Strategy for Handling Liquids

Hazard Band	Exposure Potential (EP)			
	EP1	EP2	EP3	EP4
A	Strategy 1	Strategy 1	Strategy 1	Strategy 2
B	Strategy 1	Strategy 2	Strategy 2	Strategy 3
C	Strategy 2	Strategy 3	Strategy 3	Strategy 4
D	Strategy 3	Strategy 3	Strategy 4	Strategy 4
E	Strategy 3	Strategy 4	Strategy 4	Strategy 4
F	Strategy 5	Strategy 5	Strategy 5	Strategy 5

selection, and training components are provided for each strategy. But perhaps the greatest contribution of *Containment Systems* is the whole chapter dedicated to specific examples of engineering controls that are typically used for each control band and for various scales of operation. Devices such as bag dump stations, laminar flow booths, gloveboxes, and several others are given particular attention and matched to the type of process being conducted, such as weighing and dispensing. When matched with the additional information on administrative controls and PPE, the level of guidance provided by *Containment Systems* is much more useful and pertinent to the pharmaceutical industry than that provided by COSHH Essentials or any other banding system.

Most pharmaceutical companies have adopted some variation of the level of guidance provided by this landmark work. Indeed, the majority of them have taken the initial work and over time grown and evolved the amount of guidance provided by the banding system. For instance, many organizations will include criteria on building design, material construction, workflow direction requirements, acceptability of scale, lighting requirements, noise requirements, and many more. The key output that is the greatest feature is the list of recommended of controls for any particular hazard band. This guidance significantly reduces the amount of research that is required by the industrial hygienist and focuses his or her efforts toward reducing the identified risk of an operation.

5.3   DRAWBACKS TO CONTROL BANDING

While we have spent many pages espousing the numerous advantages of control banding, it is critically important to point out the drawbacks associated with the methodology as well. No single system is perfect, and control banding is not immune from this reality.

Perhaps the greatest drawback is that personnel are often tempted to use control banding as an exposure assessment tool. That is, they will utilize the qualitative methods previously mentioned to conduct an "exposure assessment", and thereby perform the risk assessment in lieu of actual sampling. The primary problem with this approach is that any type of control banding methodology is not intended to be a predictor of occupational exposure.[4] This issue has been identified before, and yet personnel who are not fully trained in industrial hygiene utilize control banding schemes as a means to perform the "exposure assessment" of an operation in a qualitative manner.

An example can illustrate this point. Suppose a plant safety specialist is performing a "risk assessment" of an operation in which an employee charges a material into a V-blender as part of a granulation process. The substance is classified as Hazard Band B material (from Table 5.4) and is judged to be a "medium" dustiness level. Furthermore, the material is emptied from bags, each weighing 25 kg. Depending on the batch size, upward of 20 bags can be emptied for the process, taking a total of 65 minutes. Based on these criteria, the safety specialist follows the control banding workflow and assigns the Exposure Potential as EP3 from Table 5.3. The issue now arises: What exactly does EP3 mean? Since there are no units associated, it is not possible to gauge what the employee's exposure is. Furthermore, the output is listed as an "exposure potential", not an "exposure guarantee". Yet when users perform a qualitative "risk assessment" using such a methodology, the result is that the existing controls for the operation will be assessed against those recommended by the control band, and if they are not met then the assumption is that the employee is overexposed. Significant investments and retrofits will probably need to be performed as a result. However, this all assumes that the operator is indeed overexposed to the material from nothing more than the output of a qualitative risk matrix (see Chapter 4 for a more detailed discussion on the shortcomings of risk matrices). It is entirely possible that the operation is well-controlled, and the employee is not at any appreciable risk to overexposure. Quantitative sampling would provide the answer to this question but is not addressed in the qualitative risk assessment approach. It is recognized that very few industries have full-time access to industrial hygiene sampling expertise, especially in-house capabilities (the very reason control banding was developed in the first place). The pharmaceutical industry does have the resources to perform such evaluations. Ignoring such a critical step in favor of quick, qualitative assessments that lead to vague or inaccurate conclusions is a poor utilization of resources.

Another drawback lies with the assignment of substances to the specified hazard bands. The development of a system by which virtually anyone can assign chemicals to a toxicologically defined exposure band is an extremely useful tool; however, many dedicated toxicologists caution using an approach that relies upon the use of H-phrases obtained from SDSs. One reason for this is that while H-phrases are derived indirectly from toxicology data, the assignment is not based directly on the data itself. There is a wide margin of error which cannot be accounted for in such a system. In Chapter 2, we mentioned that one of the critical aspects of a toxicological evaluation is the quality of the data. By simply utilizing H-phrases from an SDS, the end user is at the mercy of the SDS author in hoping that quality studies were utilized for the assignment and that, ultimately, the appropriate H-phrase was assigned to the chemical of interest. This particular problem has also been pointed out in the past as a significant shortcoming for control banding.[3]

Compounding this issue is the anticipated exposure range itself. Assuming a substance is placed into a band appropriately, how does the practicing hygienist conduct quantitative sampling to confirm the exposure profile for an SEG? For instance, if we continue with our Hazard Band B example from earlier, the allowable exposure range for such a substance is 100–$1,000$ $\mu g/m^3$. That is a very wide range of allowable exposures. Without a definitive OEL to use as a basis of comparison, a common practice is to take several air samples and conduct traditional statistics to evaluate the $UTL_{95,95}$ of the data set. If the $UTL_{95,95}$ falls within the exposure range, the exposure is considered acceptable. But this approach requires more samples to obtain a useable data set, otherwise the hygienist is at the mercy of small sample sizes (see Chapter 4). Furthermore, how does this become classified in the modern paradigm of aligning exposure categories to the AIHA methodology? In essence, there are now two systems for classifying exposures, which lead to confusion.

Another drawback to control banding is the absence of engineering-specific information for engineering controls. This may seem puzzling, but most control banding schemes are designed to provide general guidance on controls.[5] For instance, COSHH Essentials recommends only four groups of controls: general ventilation, local exhaust ventilation, containment, and specialist advice. While a general example for each category is certainly provided, these do not provide task-specific or even industry-specific guidance on the type of control to install. This is intentional for smaller enterprises, as it would be impossible to provide specific examples for every industry and every type of task.

But again, given the unique nature of the materials and processes within the pharmaceutical industry, simple categorization is not enough. Having a thorough understanding of the advantages and disadvantages as well as variations of each class of control is paramount for the industrial hygienist. For example, having a control band recommendation of "local exhaust ventilation" is a step in the right direction, but without fully understanding the options and how they work, the hygienist and project engineer are likely to install the cheapest or most readily available LEV option without fully considering if that is appropriate or even designed correctly. The same notion applies to containment. When most engineers and industrial hygienists hear the term "containment", the immediate conclusion is to put the operation inside a glovebox. This may have been the appropriate option 20 years ago and is certainly the most prudent option for potent compounds, yet the concept of containment has evolved over time and no longer strictly requires the "glovebox only" mentality. A full and thorough understanding of these concepts is absolutely vital to ensuring that not only is the appropriate engineering control selected for the material and task, but that it is also installed correctly with the proper utility requirements (CFM, duct size, transport velocity, etc.). If the equipment is not designed and installed appropriately, it will not perform as intended and do little to enhance worker protection, giving a false sense of security.

Indeed, the mere installation of a piece of equipment from a recommended list of controls does not guarantee controlled levels of exposure. Efforts to validate control banding schemes by performing sampling after installation of recommended controls have been performed.[6] By and large, the reported results were positive in that the selected controls "reduced exposures", and in some cases claimed, the installed controls resulted in "over-controlled" scenarios. It is not entirely clear what "over-controlled" means without a point of reference, but it also begs the question regarding efficacy of the control if a quantitative starting point is not provided for comparison purposes. In the modern world of pharmaceutical industrial hygiene, the risk-based methodology is aligned with the AIHA exposure categories, so knowing how low exposures are reduced is paramount. Additionally, not all validation studies on generalized control banding methods have been reported to be positive, indicating that a one-size-fits-all approach via generalized control selection may not be as effective as once thought. Essentially, if controls are installed via generalized recommendation from a control band scheme, one should not simply assume they are effective and quantitative sampling should always be performed to validate their effectiveness.

5.4 PHARMACEUTICAL INDUSTRY WORKFLOW

The modern workflow in the pharmaceutical industry has made an effort to integrate control banding into their risk treatment and decision-making methodologies. But the modern control banding scheme has further evolved and now also considers process types in addition to material characteristics and duration. For instance, a tableting process and a bulk powder dispensing process may utilize very similar materials with identical "dustiness" levels and on the same scale, but they are inherently very different and require significantly different equipment, layouts, etc. Contemporary control banding now consists of crafting detailed repositories of controls that are subdivided and assigned to not only chemical hazard bands but also the type of task and scale of concern. More stringent types of controls are classified in higher tiered exposure bands based on the level of control they offer. In this manner, the modern control banding schemes are an amalgamation of the original PB-ECL concept pioneered by MSD and the detailed listing of control options that was put forth by Hirst, Brocklebank, and Ryder.

The typical workflow consists of hygienists performing routine or baseline sampling as part of the initial risk assessment steps, and then using the output of the risk characterization from Bayesian analysis to determine the acceptability of the risk level. If it is unacceptable, the normal response is not to automatically install new controls; rather, he or she then utilizes the control banding chart of recommended controls to perform a root cause analysis. The control banding chart, which is essentially now a compendium of requirements for each hazard band, can be used as a checklist for the hygienist. By going through the list, he or she can determine what controls, if any, are missing for the given task and type of material used. The control banding chart does not pertain to only engineering controls, either; rather, just as the PB-ECL format did, the contemporary control banding charts outline criteria for general ventilation, training requirements, PPE selection, room layout, material directional flow, security, and other items.

The inclusion of a significantly detailed list of controls for a given control band creates a checklist for the practicing hygienist against which he or she can compare if sampling results are not favorable. By knowing the control band into which the sampled material resides, the hygienist can relatively quickly assess if any crucial controls are missing or not properly followed. Reviewing the list of "required" engineering controls can often pinpoint a likely missing control that should be added or perhaps updated; in contrast, if all the appropriate controls are installed, then the source of exposure is likely due to poorly functioning engineering controls or inadequate administrative controls. In these cases, a root cause analysis (RCA) is more suitable to identifying the source of exposure rather than a CAPEX project to correct the issue. Additionally, the control bands provide a terrific blueprint when new labs or facilities are being designed from scratch. This particular application of control banding saves significant time and money by ensuring all the required equipment and utilities are in place for a desired process. It is far better and less expensive to install everything up front rather undergo retrofits later.

These applications are the most common utilization of control banding in the pharmaceutical industry. The banding system enables the practicing hygienist to ascertain if appropriate and specific controls are in place for a given process or particular hazardous material. The inclusion of various types of engineering controls and required administrative controls is based off years of industrial hygiene evaluations of countless variations, enabling a thorough understanding of the advantages and limitations of various classes of equipment. At the time of this writing there was not a single, universally accepted control banding scheme for the pharmaceutical industry. Each organization has developed its own based on years of experience and collaboration with external experts. Furthermore, the accepted types of controls that are often recommended for each band are based on not just acceptable risk criteria such as containment level, but also on important supporting criteria such as ergonomic compatibility and energy requirements. We will see in later chapters how sustainability is a major driving force in designing controls for processes.

It is worth noting that some organizations have not chosen to disregard the qualitative risk assessment motif from control banding. Indeed, these organizations often utilize the "exposure potential" paradigm as a first-pass risk assessment for prioritizing detailed sampling endeavors at a later date. This is particularly true for sites which have a very large number of processes and materials. Such a strategy represents a unique departure from other previously discussed qualitative risk assessments in that the output of the process is not used for risk-based decision making; that is, a definitive verdict of an exposure profile is not being pronounced. Instead, it is used as a means to decide which detailed sampling plans to design and execute first. It is not uncommon for detailed quantitative sampling analysis of a process to later show that an exposure profile is drastically different from the qualitative output. But the quantitative analysis is what is used for the risk-based decision making, not the qualitative assessment exercise.

5.4.1 Understanding the Material

The implementation of control banding as a tool for guiding industrial hygienists and other risk managers toward selecting an appropriate engineering control based on the hazard category is a massive step forward in risk reduction. But simply following a control banding chart in a "plug-and-chug" format can potentially be detrimental. Fully understanding the properties of the material being handled in a given process is essential to proper control.

5.4.1.1 Particle Size Distribution

One of the most important physical characteristics of any solid material is the particle size distribution. The size of the particles affects various other properties that are important in the pharmaceutical sector, such as solubility and compressibility. Powders and other solids used in the pharmaceutical industry all consist of particles of varying sizes, often from sub-micron diameters to several hundred-micron diameters. Knowing what the particle size distribution is for a given substance can provide significant insight, and the particle size distribution can provide such information.

From an industrial hygiene perspective, particle size is important because it closely correlates to how dusty a solid substance can be. Dustiness of powders has been a major topic within the pharmaceutical industry for decades, and many efforts have been made to standardize ways to measure dustiness.[7] However, one of the best metrics for predicting powder dustiness is the particle size distribution of the material.[8] Generally speaking, the smaller the particle diameter, the dustier a substance becomes. This makes some logical sense, since materials with smaller diameters take longer to settle out of the air and are therefore more available for inhalation. By looking at the particle size distribution and gauging where various percentiles of the sizes are, one can roughly gauge the dustiness of a material and how much of a risk it poses to the worker. Concerning dusts and particulates, industrial hygienists are often concerned with those materials possessing diameters <100 μm, as this represents the inhalable fraction of powders. Substances whose diameters are in this range have the potential for employee exposures.

But aside from allowing the hygienist to predict dustiness and exposure potential, particle size also has a significant impact on control selection. Effective control relies on moving airborne contaminants away from the worker, usually in an airstream. Generally speaking, smaller diameter particles are easier to capture and transport in air than their larger diameter counterparts, especially if the airborne material is generated with significant energy. This is entirely due to physics and overcoming the larger particles' inertia with an air current. Smaller particles simply have less mass and less inertia, requiring less energy to change their vector. This is the entire reason why inertial impactors can separate particulates by size and cyclones can be used to sample for specific particulate diameters, but the same principle applies to capturing unwanted airborne contaminants of various diameters. Therefore, by knowing what the particle size distribution is for a powder or a granulated material, the hygienist can make informed recommendations regarding the capture velocity, capture distance, volumetric airflow rate, and transport velocity (see the section on Local Exhaust Ventilation later in this chapter).

An important consideration is not just the particle size distribution for a given raw material, but rather how that distribution changes as it undergoes various uses within a pharmaceutical process. For instance, a common process in the pharmaceutical industry is milling, a process in which energy is applied to coarse, large particles and breaks them down into smaller ones. In other words, solid materials are mechanically "broken up" to alter the particle size distribution to decrease larger diameter particles and increase smaller diameter ones. Milling is routinely applied to final products but can be applied elsewhere in a manufacturing lifecycle as well. In direct contrast, the process known as granulation creates larger particles from smaller ones through agglomeration and other binding techniques. The products from granulation are often significantly less dusty and free-flowing, properties that are inherently desirable during downstream compaction processes. For the practicing hygienist, knowing how the particle size changes can significantly alter the recommended controls for an operation. The required utilities can be significantly different than initially thought, and if not designed for appropriate size distribution, then an expensive and sub-optimal control will be the result.

5.4.1.2 Combustibility

Particle size distribution is not the only physical property or characteristic that is of importance when selecting an appropriate control. One of the most overlooked properties of solid materials when selecting a control is the potential for combustibility. Similar in nature to flammable vapors, combustible powders differ in that they also require significant dispersion into the air and also require tight spaces (confinement) in addition to the variables of oxygen, ignition source, and fuel of the fire triangle. The result is the combustible dust pentagon (Figure 5.4). Such conditions are commonly encountered within ventilation ductwork, making dust combustibility a significant concern during ventilation system design. There are numerous substances used in the pharmaceutical industry that are combustible solids, including adipic acid, magnesium stearate, dextrose, lactose, mannitol, sorbitol, and sucrose. Unsurprisingly, all of these materials are excipients and are often used in very large quantities in secondary pharmaceutical sites.

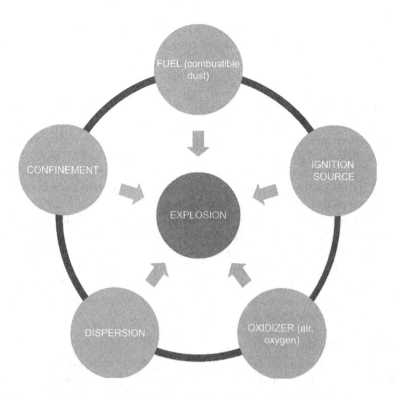

FIGURE 5.4 The combustible powder pentagon adds the variables of powder dispersion and confinement to the well-known fire triangle.

For any combustible dust, there are two key parameters that every hygienist should know: minimum ignition energy (MIE) and minimum explosive concentration (MEC). The MIE is the lowest amount of energy needed in a spark (such as through electrostatic discharge from operators or equipment) that is needed to ignite the solid in a dust cloud mixture. Numerous companies are able to perform MIE assessments in accordance with standard testing methods such as ASTM E2019. The MEC is the lowest concentration necessary for a dust cloud to support deflagration, or the creation of a fireball. The MEC is analogous to the lower explosive limit (LEL) for a flammable gas or vapor. Just as the MIE can be tested for according to national and international standard, the MEC can also be tested according to AST E2931. While these test methods are available and several outlets are able to perform them in ISO 17025 accredited labs, such information can be unavailable for relatively uncommon or new materials.

It is worth noting that dust combustibility represents a physical hazard, and one that normally falls into the arena of process safety management (PSM) specialists. While this is entirely true, industrial hygienists within the pharmaceutical sector still probe into these matters by consulting with PSM experts on all risk remediation projects. The reason for this is because if a combustible material being controlled requires transport via ventilation, additional regulatory standards can potentially apply which impacts the selection and design of those controls. For instance, NFPA 652 states that a combustible dust is any particulate matter which has an aerodynamic diameter of 500 µm or less. This diameter was chosen because particles with larger diameters than 500 µm often do not have enough surface-to-volume ratio and do not support the deflagration process very well. Once again, to help ascertain the answer if a bulk material meets the definition of combustible dust, the practicing hygienist can turn to the particle size distribution. The distribution will easily provide the answer as to how much of the material meets the definition of combustible dust size. However, there are no specific requirements as to how much of the material must meet the size requirement. It is incumbent on the PSM specialist to determine, based on the particle size distribution of the material and the quantity being handled in a given process, whether or not the required size criteria would meet the MEC in a process. Such a determination is extremely technical and should be handled only by PSM experts.

But there is a more important factor for combustible dusts that directly impacts the industrial hygienist. If a material is indeed classified as a combustible dust and it is being controlled by some form of ventilation, then there are minimum transport velocities that must be met. While there are no regulations dictating air transport velocities within ventilation systems for combustible dusts specifically in the pharmaceutical industry, there are general industry recommendations. NFPA 91, *Standard for Exhaust Systems for Air Conveying of Vapors, Gases, Mists, and Particulate Solids*, states very general criteria for ensuring that air cleaning devices are properly designed to ensure that they convey (move) all contaminants throughout the system without resulting in material buildup.[9] The standard, in turn, references industry best-practices for material conveyance. Such recommendations come from authoritative sources such as the ACGIH and are largely based on the density and particle size of the material being conveyed.[10] Consequently, recommended flow rates through ventilation systems are not dictated strictly by the physical or health hazards but on the intrinsic physical properties of the material. Again, it is incumbent on the hygienist to recognize these potential pitfalls and to design the control system appropriately so as to not create a potential life-threatening situation within the system (we will go into greater depth on ventilation transport rates for ventilation systems in Chapter 6).

5.4.1.3   Understanding the Process

In Chapter 3, we went into detail on how the practicing hygienist must understand a given process in order to design a sampling plan to identify likely points of employee exposure. In such a circumstance, we are attempting to understand how the equipment and material flow occurs as it currently exists. When the hygienist is attempting to install new controls or design a process from the ground up, he or she must be concerned with not just potential exposure control, but how the intended control will affect the process holistically.

Perhaps the biggest consideration that must be realized is that any control, no matter how small or large, will impact the timing of the process; that is, there will be some effect as to how long a particular process takes to run from start to finish and produce a product. A consistently recurring issue, primarily among non-hygienists involved in the risk management process, is the assumption that an identified control can be identified and installed on a process, and there will be absolutely no effect on the process output. This misconception is one which should be addressed upfront whenever a control is identified so there will be no surprises for all stakeholders. Unfortunately, it is often difficult (if not outright impossible) to predict just how much of an impact there will be on production. Despite this limitation, it is imperative that the hygienist and engineering team convey to the production team that changes to the production schedule can and should be anticipated as a result of the impending change, which is necessary from a safety standpoint. This remains one of the shining examples for modern organizations that place employee safety over production efficiency.

In addition to impacts on production schedules, there are other downstream effects that should also be taken into consideration. One of these are cleaning requirements, especially if the control is to be used within a GMP environment wherein change overs occur. All equipment must be cleaned periodically to maintain good working order and to prevent cross contamination, and industrial hygiene controls are no different. For many pharmaceutical processes, in particular enclosures in which potent compounds are handled, unique features such as clean in place (CIP) are routinely offered as a convenience. Many CIP protocols will utilize a detergent or other substance to facilitate the cleaning process, but the introduction of another chemical into an environment where other substances which will be ingested and/or consumed by patients and consumers will be handled introduces a significant quality variable. The risk of cross contamination of product is a significant concern as it has massive downstream impacts. Cross contamination can potentially result in product recalls or worse, in affecting patient health. Such concerns are not strictly relegated to cleaning regimens, but they are an obvious source of cross contamination and one which can and should be addressed at the outset of control selection. This is why at least one member of the Risk Management Team should be a member of the Quality Department.

It is worth noting that CIP and enclosures are not the only way for cleaning to have an impact on control design and selection. Common LEV and dust collector systems all work to divert contaminants away from employees, but the dust collectors and the ductwork through which the material is transported should all be cleaned on a regular basis. For improperly designed ventilation systems, insufficient transport velocities lead to significant deposition of particulate, which can pose a physical hazard (see the previous section on Combustibility). The ductwork needs to be cleaned (or at least inspected) on a routine basis, and yet the very process of doing so puts the individuals performing said cleaning task at risk of exposure. In effect, it creates a whole new SEG which the hygienist must take into consideration. Proper design of such systems can drastically reduce exposures during cleaning, but these must be considered during the design phase to avoid costly retrofits later. Fortunately, dust collectors have evolved over the years to incorporate safer ways of emptying them and changing filters, but such enhanced safety systems often increase the cost of the unit.

5.5 SUMMARY

In situations where there is an appreciable risk of exposure to employees in the workplace, a common means of risk treatment is to reduce the risk. Risk reduction is often accomplished by implementing solutions from the hierarchy of controls. The hierarchy of controls stipulates that engineering controls are the most effective and practical means of reducing employee exposures. In the past, industrial hygienists might have simply provided general advice for engineering controls, such as "install LEV", but contemporary pharmaceutical industrial hygienists are now expected to be a fully integrated team member in the selection of engineering controls. Control banding schemes are frequently utilized in the pharmaceutical industry to aid in selecting the appropriate level of engineering control. Older control banding schemes, or "performance-based exposure control levels", provided groundbreaking guidance when they were first introduced. The schemes have evolved over time and can now contain very detailed guidance, including the types of equipment to install and the desired amount of utilities for the controls (airflow, transport velocity, etc.). To fully make an informed decision, the IH must fully understand the process and the material being handled. Aspects such as particle size, combustibility, and cGMP requirements must be considered when choosing a control that fits into the required banding scheme.

NOTES

1 1) Farris, J. P., Ader, A. W., and Ku, R. H. (2006). History, implementation and evolution of the pharmaceutical hazard categorization and control system. *Chimica Oggi - Chemistry Today, 24*(2), 5–10. 2) Zalk, D. M. and Nelson, D. I. (2008). History and evolution of control banding: A review. *Journal of Occupational and Environmental Hygiene, 5*(5), 330–346.

2 Naumann, B. D., Sargent, E. V., Starkman, B. S., Fraser, W. J., Becker, G. T., and Kirk, G. D. (1996). Performance-based exposure control limits for pharmaceutical active ingredients. *American Industrial Hygiene Association Journal, 57*(1), 33–42.

3 Hirst, N., Brocklebank, M., and Ryder, M. (Eds.). (2002). *Containment Systems: A Design Guide.* IChemE.

4 Evans, P. and Garrod, A. (2006). Evaluation of COSHH essentials for vapour degreasing and bag-filling operations. *Annals of Occupational Hygiene, 50*(6), 641.

5 Jones, R. M. and Nicas, M. (2006). Reply: Letter to the editor: Evaluation of the utility and reliability of COSHH essentials. *Annals of Occupational Hygiene, 50*(6), 643–644.

6 1) Tischer, M., Bredendiek-Kamper, S., and Poppek, U. (2003). Evaluation of the HSE COSHH essentials exposure predictive model on the basis of the BAuA field studies and existing substances exposure data. *Annals of Occupational Hygiene, 47*(7), 557–569. 2) Hashimoto, H., et al. (2007). Evaluation of the control banding method - comparison with measurement-based comprehensive risk assessment. *Journal of Occupational Health, 49*(6), 482–492.

7 1) Cawley, B. and Leith, D. (1993). Bench-top apparatus to examine factors that affect dust genera-tion. *Applied Occupational and Environmental Hygiene, 8*(7), 624–631. 2) Cowherd, C., et al. (1989). An apparatus and methodology for predicting the dustiness of materials. *American Industrial Hygiene Association Journal, 50*(3), 123–130. 3) Plinke, M. A., Maus, R., and Leith, D. (1992). Experimental examination of factors that affect dust generation by using Heubach and MRI testers. *American Industrial Hygiene Association Journal, 53*(5), 325–330. 4) Plinke, M. A., Leith, D., Boundy, M. G., and Loffler, F. (1995). Dust generation from handling powders in industry. *American Industrial Hygiene Association Journal, 56*(3), 251–257.

8 1) Pujara, C. P. and Kildsig, D. O. (2001). Effect of Individual Particle Characteristics on Airborne Emissions. In Wood, J. (Ed.), *Containment in the Pharmaceutical Industry*, Marcel Dekker Publishing. 2) Schofield, C., Sutton, H. M., and Waters, K. A. N. (1979). The generation of dust by materials handling operations. *Journal of Powder Bulk Solids Technology, 3*(1), 40. 3) Lopez, L., et al. (2017). Particle size distribution: A key factor in estimating powder dustiness. *Journal of Occupational and Environmental Hygiene, 14*(2), 975–985.

9 NFPA 91–2019. "Standard for exhaust systems for air conveying of vapors, gases, mists, and particulate solids".

10 ACGIH. *Industrial Ventilation: A Manual of Recommended Practice for Design*, 30th Ed. 2021.

6 Risk Treatment
Local Exhaust Ventilation (LEV)

6.1 INTRODUCTION

The choice to reduce an identified unacceptable exposure risk to employees is often followed by deliberation as to how the risk reduction should be accomplished. More often than not, the risk management team leans on the industrial hygienist to recommend a particular control to achieve the desired goal (Figure 6.1). Historically, this was where the involvement for the hygienist ended, as the selected control type was then given to an engineering team (sometimes an external third party with little to no stake in the outcome) who then designed and built the recommended control. Two significant issues frequently arose from this workflow. First, there was not always a clear directive as to what level of airborne material the device should perform to; that is, how well would the device work and what would the anticipated airborne exposures be post installation? Second, organizations would often take the recommendation from the industrial hygienist and then proceed to install the least expensive option available, taking as many shortcuts as possible.

Thankfully, this is no longer the expected workflow in the ISO 31000 paradigm. Especially in the world of pharmaceuticals, the hygienist plays a critical and dynamic role at the risk treatment stage. Not only is he or she expected to make the recommendation of the type of control given the inherent

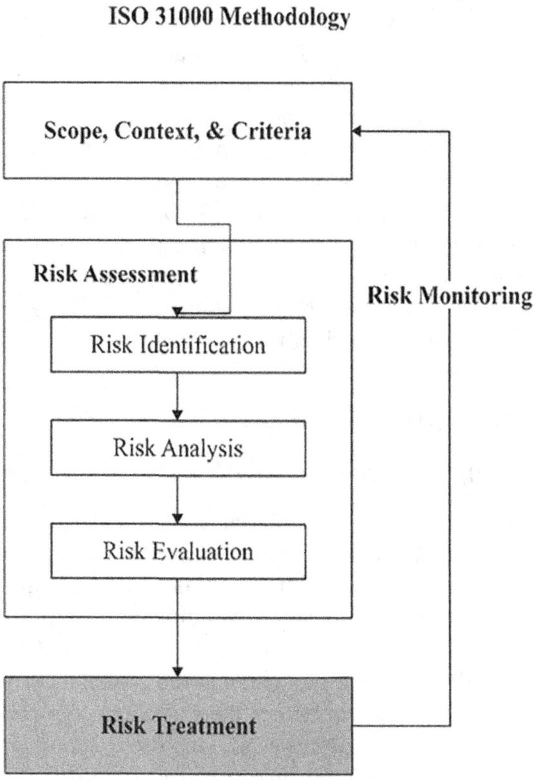

ISO 31000 Methodology

FIGURE 6.1   One of the final steps of the ISO 31000 risk management flow is risk treatment.

DOI: 10.1201/9781003273455-6

health hazard posed by the material but provide insight as to its design and expected performance. The principles outlined in the previous chapter provide significant guidance to the hygienist in successfully implementing the first part of those goals (selection of a control). A detailed control banding scheme can point the hygienist in the right direction as to the general types of engineering controls to choose based on the exposure band in which the material(s) fall. Also understanding the process (workflow, process changes to the material, timing, production demands, anticipated changeovers, etc.) as well as the physical characteristics of the material (particle size, density, combustibility) will provide additional insight for control selection.

However, these do not necessarily provide guidance on how to design the control so it provides the anticipated level of protection. Perhaps just as importantly in the modern era, an improperly designed control will not work as efficiently as it possibly could. In this regard, additional utilities (i.e., electricity) are required to make a control operate at optimal performance. Energy efficiency and sustainability are a major driver in the decision-making process among 21st-century pharmaceutical organizations. Consequently, poorly designed controls will receive additional scrutiny and are easy candidates for expensive retrofitting. It is far better to spend a little more money upfront to initially design a control appropriately than to spend several times the original amount in retrofitting. This is one of the most underrated ways that the modern industrial hygienist can provide significant value and impact to a site – not just by lowering exposure risk to personnel, but by doing so in an efficient and as sustainable manner as possible.

Local exhaust ventilation (LEV) has been one of the most common forms of engineering control over the last 200 years. The multitude of options and relative ease of installation made it a popular choice in general industry for controlling workplace emissions of contaminants. LEV remains a common choice in the pharmaceutical industry as well. A common mantra regarding these systems is that the purpose of LEV is to remove as much of the contaminant as possible with as little air as possible while using the least amount of electricity to accomplish the feat. This chapter will review the basic components of a LEV system and address some pitfalls that are routinely found among installed systems. Chapter 7 will take a much deeper look at the concept of containment and various hoods and how they impact the operator and the operation.

This chapter is not intended to be a significant deep dive into all aspects of ventilation. Rather, it is intended to be a review of key features that are of importance to the practicing industrial hygienist. For a more thorough review of the theories and applications of ventilation, the reader is encouraged to read through the ACGIH publication *Industrial Ventilation: A Manual of Recommended Practice for Design, 30th Edition*.[1] It is often considered the cornerstone reference for industrial ventilation and a copy should be on the bookshelf of every practicing industrial hygienist. Additionally, many of the same principles are utilized in the voluntary standard ANSI/ASSP Z9.2–2018. It is highly advised that the reader be familiar with both of these documents.

6.2   A REVIEW OF AIR FLOW PRINCIPLES

6.2.1   Static Pressure (SP), Velocity Pressure (VP), and Total Pressure (TP)

Before reviewing key details of LEV systems, it is prudent to understand some principles of air and airflow (Figure 6.2). The first concept is that within exhaust systems, there are several different types of air pressures which must be accounted for. These are static pressure (SP), velocity pressure (VP), and total pressure (TP). SP is the pressure exerted in all direction within a duct, and within exhaust systems attempts to collapse a duct (negative static pressure) or expand/press upon the walls of the duct (positive static pressure). SP can also be thought of as the potential energy of the system, and it is the force that is used to overcome airflow resistance (filters, friction, contractions, bends, elbows, etc.).

In contrast, VP is the kinetic energy that moves air (and the contaminants within it) from rest to a desired velocity. Unlike SP, which can be positive or negative in value, VP can only be positive. It is directly related to air velocity by the equation:

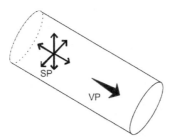

FIGURE 6.2   A diagram showcasing SP, VP, and TP within a LEV system.

$$\text{Velocity (V)} = 4{,}005\sqrt{\frac{\text{VP}}{\text{df}}} \tag{6.1}$$

In Equation 6.1, velocity (V) is given in units of feet per minute (ft/min), VP has units of inches of water column (in wc), and df is the density factor (see the next section for a brief discussion on density factor). Importantly, Equation 6.1 applies to the use of imperial units and a different variation can be used for SI units.

Total pressure (TP) is the third and final form of pressure within a LEV system. It is equal to the sum of the SP and the VP within the system:

$$\text{Total Pressure (TP)} = \text{SP} + \text{VP} \tag{6.2}$$

Of particular importance to the industrial hygienist is how to measure these variables within the LEV system. SP can be measured directly using a manometer and is typically reported in the units of inches of water column (in. wc). In contrast, VP cannot be directly measured; rather, it is measured indirectly using a pitot tube which measures SP and TP simultaneously. The hygienist can then solve for VP algebraically by rearranging Equation 6.2. Knowing what these values represent and how to measure them are vital to not only designing the LEV system (as we will see later) but also diagnosing any potential problems with the system.

6.2.2   Air Density and Density Factor

A second principle of air for the practicing hygienist to keep in mind is that air has density. We may not always realize it, but air possesses a specific mass per unit volume. This concept is important because the density of air directly relates to how much air is being drawn through a system, which in turn has a direct effect on velocity. At standard conditions, air has a density of 0.075 lb/ft^3. In ventilation, standard conditions are defined as 70°F and 1 atmosphere (atm) of ambient air pressure. The vast majority of ventilation systems are designed with the assumption that standard conditions will be met, so no deviation from standard values is utilized. However, there are several factors that directly impact density, and these are summed in a single variable called the density factor (*df*, Equation 6.3).

$$df = \left(df_e\right) \times \left(df_p\right) \times \left(df_t\right) \times \left(df_m\right) \tag{6.3}$$

Each variable that comprises *df* must be calculated individually. These are summarized in Table 6.1. Df_e relates to elevation. At higher elevations, the air is thinner and thus is less dense. It is directly correlated to the number of feet above sea level (*z*). Df_p is the pressure exerted within the duct, but not to be confused with TP. Df_p is solved for by adding the SP within the duct to 407, the number of inches of water column exerted by standard air. The temperature variable, df_t, is important because warmer air expands and thus impacts the total volume being moved. Df_t is solved for by adding the

TABLE 6.1
Density Factor Individual Components[1]

Term	Equation	
Elevation factor (df_e)	$df_e = \left[1 - \left((6.73)\left(10^{-6}\right)(z)\right)\right]^{5.258}$	Equation 6.4
Pressure factor (df_p)	$df_p = \left[407 + (SP)\right] \big/ 407$	Equation 6.5
Temperature factor (df_t)	$df_t = 530 \big/ (T + 460)$	Equation 6.6
Moisture factor (df_m)	$df_m = (1 + \omega) \big/ (1 + 1.607\omega)$	Equation 6.7

temperature of the incoming air, in °F, to 460 to obtain the absolute temperature in degrees Rankin. Finally, the moisture factor, dfm, is found by using Equation 6.7. For the moisture factor, ω refers to the humidity ratio, which is the mass of water in the air per pound of air.

Each of these aspects should be accounted for when designing an LEV system. In ventilation systems, non-standard conditions are considered to be met if df deviates by ±5%. Such a condition can easily be met if each input variable alters by as little as 1.3%. We have already encountered where df can be used (Equation 6.1). When air is less dense (a $df < 1$), air moves faster, which in turn equates to a larger volume of air in total. For instance, if the VP of in a duct is measured to be 0.5 in. w.c., at standard conditions the velocity is 2,832 ft/min, but if the density factor is 0.95 (a deviation of just 5%), the speed becomes 2,905 ft/min. This may not seem like a drastic difference but could have significant impact in downstream operations (as we will see later).

6.2.3 MAKEUP AIR

One of the most frequently overlooked components during LEV design, makeup air is a critical component to making sure the building is balanced. The air that enters an LEV system must go somewhere. The majority of systems used in the pharmaceutical system are exhausted outside the building to the general environment. If we think about all the air that these systems exhaust, that same volume of air must be replaced (or very close to it). Without replacing the exhausted air, the building in which the LEV system is placed will become extremely negatively pressured, resulting in rogue cross-drafts and doors that either leak air or are impossible to open. Such scenarios create additional unwanted safety hazards.

To prevent this, makeup air systems must be installed alongside the LEV system. Makeup air systems are not special or unique components; rather, they are typical HVAC units. These HVAC units must be sized appropriately for the area under consideration, and the ductwork must be run to provide the makeup air in question. But the real cost is associated with how much air the exhaust and makeup air systems are utilizing. In essence, the HVAC unit will be supplying conditioned air (i.e., air that is heated, cooled, and/or moistened or dried) to an area that will simply be exhausting said conditioned air. These conditions are referred to as single-pass air (or 100% outside air) since there is no recirculation of conditioned air. In routine indoor work environments, a significant percentage of indoor air is recirculated to save money on heating and cooling costs. These cost saving efforts are not usually applicable in a pharmaceutical site due to the potential of recirculating airborne contaminants. Consequently, there is a hefty sum of money invested in single-pass makeup air. Selection of a proper LEV system which abides by the previously stated mantra (remove as much of the contaminant as possible with as little air as possible) is key to minimizing this impact. The conscientious industrial hygienist and engineer on the risk management team cannot ignore the cost and environmental impact of makeup air on poorly designed LEV systems.

6.3 LEV SYSTEM COMPONENTS

LEV systems are ubiquitous in general industry and at pharmaceutical sites as well. Having a firm understanding of all the components of an LEV system is of paramount importance for the practicing industrial hygienist. Such an understanding aids in the design of the system not only to ensure proper risk reduction is obtained but also to maintain the system and diagnose any potential problems that may arise.

Every LEV system is comprised of five components. In the order in which they are typically encountered, these are the hood, the ductwork, the air-cleaning device, the fan, and the exhaust stack. An overly simplified LEV system depicting these components is shown in Figure 6.3. Each one of these components must be designed appropriately to achieve the anticipated level of contaminant capture and to minimize utility consumption.

6.3.1 HOODS

When most people think of the term "hood", they immediately conjure an image of the commonly encountered chemical fume hood (also called a chemical cabinet in countries outside the United States). However, in LEV systems, a hood is simply any portal or entry into which air flows. Hoods are where the entire system begins and as such should receive a significant amount of attention during the selection and design stage.

There are dozens of LEV hood designs available on the market, and many are specially designed for a specific industry or task. In the pharmaceutical industry, a handful of designs are routinely used with modifications fit for purpose. Importantly, since such designs have been used extensively throughout pharmaceutical organizations, the anticipated containment level provided by the hoods is roughly known. Generally speaking, the industrial hygienist can anticipate a typical LEV hood to provide powder and dust exposure control anywhere in the range of 500 µg/m³ to as low as 50 µg/m³, perhaps lower depending on scale and ways of working. For vapors from solvents and liquids, the anticipated exposure levels can be as low as 1 ppm. However, it is vital to understand that such airborne contaminant levels are dependent on many factors, including the amount of material being handled in the operation, the volatility/particle size of the material, process temperature, and energy of the operation. All of these factors were discussed in the previous chapter as significant aspects to understand prior to choosing a control.

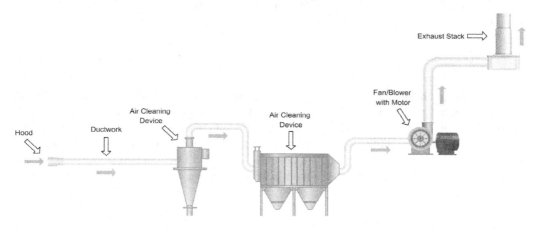

FIGURE 6.3   A simple LEV schematic showing the hood, ductwork, air-cleaning device, fan, and exhaust stack.

6.3.1.1   Distance from Emission Source and Capture Velocity

Yet there are other even more important factors which directly affect how a hood will perform. One of these is how much air is drawn through the hood. Suppliers of commercially available LEV hoods frequently state how many cubic feet of air per minute (CFM) are needed for the hood to work effectively. Without the appropriate volumetric flow of exhaust, it is highly unlikely that a hood will provide adequate protection of any kind for employees. To say that knowing how much exhaust is available in a system is important would be a significant understatement.

But having an appropriate amount of CFM is only one part of the LEV system effectively working. Another aspect that is often ignored, especially for retrofit projects, is the distance from the source. The distance from the hood to the emission source heavily dictates how well the capture velocity of the hood will perform. Capture velocity is the minimum air stream velocity that is needed to "capture" and move the airborne contaminant into the hood for transport throughout the exhaust system. In effect, it is the velocity needed to be exerted on the airborne materials to have them overcome their own inertia, altering their vector and trajectory so they will be moved away from the worker (knowing the particle size distribution and how the material is being handled is vital for estimating the desired capture velocity, concepts which were touched upon in Chapter 5). But capture velocity is a function of the distance between the hood and the source. The larger the distance between the two, the more air is required to achieve the desired capture velocity. An often-cited rule of thumb is that the capture velocity of a plain hood drops to 10% of the face velocity at a distance equal to 1 duct diameter. For example, if a 6-inch plain hood had a face velocity of 1,000 ft/min, the capture velocity would fall to approximately 100 ft/min at a distance of 6 inches from the hood. This estimation does not exactly hold true for all hoods but is a close enough approximation.

ACGIH has published a series of empirically derived ventilation equations that are frequently used for determining how much CFM is needed for a particular hood.[1] The input variables for these equations are often hood area (ft²), desired capture velocity (ft/min), and distance from the source (ft). However, oftentimes the situation in the pharmaceutical sector (again, especially for retrofit projects) is that a certain amount of CFM is already known, so the issue becomes how close must the hood be to the operation for the hood to work properly? The equations can be rearranged to solve for this issue to give an approximate distance requirement.

An example can demonstrate the utility of these equations. Suppose there is a weighing operation wherein employees weigh out powdered caffeine for a process within a consumer healthcare site. The project engineer finds a 4-foot-wide flanged slotted hood from a supplier that claims it can provide the desired level of airborne dust control for the operation with 750 CFM. The local EHS group recommends a capture velocity of 250 ft/min. The required working distance from the hood is approximated to be:

$$\text{Distance (ft)} = \frac{Q}{2.6 \times L \times V} = \frac{750}{2.6 \times 4 \times 250} = \frac{750}{2,600} = 0.288 \text{ ft} = 3.5 \text{ inches} \qquad (6.8)$$

From the ventilation equation, the distance from the weighing operation to the hood face should not exceed 3.5 inches. However, let's assume that the location has some additional area constraints, and that the operation has to take place 12 inches away from the hood face so as to provide minimal disruption to the weighing process. Using the same equation, we now know how far away we must operate, but now we need to tabulate how much CFM will be required for the hood at the same desired capture velocity. With the modified variables, we can see that the required CFM would now be:

$$\text{Airflow (CFM, or } Q) = 2.6 \times L \times V \times X = 2.6 \times 4 \times 250 \times 1 = 2,600 \text{ CFM} \qquad (6.9)$$

For the required parameters involving the selected hood, a volumetric airflow of 2,600 CFM would be required to provide the needed capture velocity. This amount, which is almost 3.5 times the originally quoted amount of required CFM, may not seem like a drastic increase but there are substantial downstream effects. First, a larger fan would need to be selected along with a larger motor, which would cost significantly more than anticipated (we will discuss fans and motors later). Second, the additional CFM being exhausted would require additional makeup air, which has significant sustainability costs associated with it. Third, the supplier quoted the hood capture capability at 750 CFM. It would be dubious to assume that more CFM would provide the same protection level. More airflow does not necessarily equal more or better employee protection.

In the previous example, the chosen capture velocity was arbitrarily chosen to be 250 ft/min, but in reality, the choice of capture velocity is done with significant intent. ACGIH recommends a wide range of LEV capture velocities, from as low as 75 ft/min up to 2,000 ft/min. The specific velocity is dictated by the amount of energy given to the emission operation. The rationale for this is something we have already mentioned but is worth repeating. Higher energy operations (e.g., milling, blending, material conveyance, etc.) give significant speed to the particulate matter. If the hood is positioned in such a way that it must change the trajectory of the material, then higher velocities are required. The particle size and its density also play a role in selecting the capture velocity. Taken all together, the choice of capture velocity and distance dictate the volumetric flow rate (CFM) of air needed for the LEV hood to operate. The farther the distance from the source or the higher velocity required, the more CFM will be needed. Consequently, the more CFM needed for the hood means more makeup air will be required and potentially a larger fan, which equates to larger utility bills. As it can be seen, ensuring the hood is properly designed for the process at hand is extremely important as it directly impacts other aspects to the LEV system as well.

6.3.1.2 Coefficient of Entry (Ce)

From a design perspective, knowing the distance from the source, capture velocity, and anticipated CFM are still not enough to properly design the LEV system and hood. A hood operates by capturing and conveying contaminants from the environment into the LEV system. This is accomplished by the conversion of static pressure (SP) into velocity pressure (VP). The velocity pressure, in turn, is directly related to the velocity of the air stream within the hood and ductwork (Equation 6.1). Importantly, this velocity, often referred to as the transport velocity, is not the same as capture velocity. Transport velocity occurs within the hood and ductwork, whereas the capture velocity occurs outside the hood and directs airborne contaminants into the hood. Once the contaminants enter the hood, they rapidly accelerate to the transport velocity.

Every hood ever made differs on how well it converts SP into VP. It is a function of how well the hood was engineered and constructed. No hood is able to convert all SP into VP; this is simply impossible to accomplish. There are always losses of SP within the hood, but a well-designed hood can minimize these losses. Conveniently, the efficiency with which a hood converts SP within a hood (specifically denoted as SP_h) into VP is referred to as the coefficient of entry (Ce). Mathematically, this relationship can be expressed as:

$$Ce = \sqrt{\frac{VP}{|SP_h|}} \tag{6.10}$$

In Equation 6.10, the SP_h term is utilized as the absolute value since SP taken within hoods will be negative (they are on the exhaust side of the fan). The resulting equation would yield an imaginary number, so the absolute value is utilized in the calculation. Since every hood has losses in SP_h, the Ce for a hood will always be less than one, but it should be readily apparent that the closer Ce is to 1, the more effective the hood is at converting SP_h into VP. Every manufacturer should have the Ce of their hoods on file if not freely available, and this information should unquestionably be requested by the engineering team. The utility of Equation 6.10 should also be readily apparent, as it allows the industrial hygienist and the risk management team to calculate how much SP_h would be required for a hood at a desired performance.

This can be shown by continuing with the previous example. Let's suppose that the 4-foot flanged slotted hood has a Ce equal to 0.7 (value provided by the vendor) and is connected to the LEV system via a 10-inch diameter segment of ductwork. With the previously calculated flowrate requirement of 2,600 CFM, this amounts to a transport velocity (V) of approximately 4,700 ft/min (we will touch on calculations of transport velocity in the next section). From Equation 6.1, this correlates to a VP of 1.42 in. w.c.:

$$VP = \left(\frac{4,700}{4,005} \right)^2 = 1.42 \text{ in. w.c.} \tag{6.11}$$

With the anticipated value of VP in hand, we can now rearrange Equation 6.10 to solve for SP_h:

$$|SP_h| = \frac{VP}{Ce^2} = \frac{1.42}{0.7^2} = 2.89 \text{ in.w.c.} \tag{6.12}$$

From Equation 6.12, we can see that 2.89 inches of SP_h will be required for the anticipated hood. At this SP_h, the hood should be able to provide the necessary VP, which in turn will create the needed transport velocity. The desired velocity then translates into the requisite CFM, and thus, the capture velocity based on the distance from the emission source. All of these individual components must be evaluated during the design stage in order to effectively deliver on the intention of risk reduction for the employee. Historically, in some organizations, little attention was paid to the required SP_h of a hood. If a hood was chosen that was supposed to operate at a given CFM, it might have been installed without any calculations done beforehand. Upon installation, if the anticipated airflow wasn't met, the solution would often be to "increase the fan speed", which could have disastrous consequences for multi-component hood systems (as we will see later, there are limits to what increasing and decreasing fan speed can do). Luckily, in contemporary pharmaceutical organizations, such short sightedness is rarely encountered, and a significant amount of attention is paid to these small but critically important details.

It is worth stating just how impactful the SP_h requirement is and how Ce can drastically affect the entire LEV system. Figure 6.4 shows a graphical representation of the required SP_h for hoods possessing various Ce values if they were to be substituted into our previous example problem. Our supposed hood of choice, with a Ce of 0.7, has a required SP_h of 2.89 in. w.c. (at a volumetric flow rate of 2,600 CFM through a 10-inch diameter duct). As it can be seen, as the Ce decreases (and thus the efficiency of the hood), considerably more SP_h is required to be supplied to the hood. As we mentioned before, this translates directly to higher energy costs which does not bode well in today's sustainability-driven infrastructure. It is worthwhile to invest a little more money in a well-designed hood that has a higher Ce to save money in the long run. The company is more likely to achieve a faster return on investment (ROI) with this approach. Fortunately, most commercially available hoods which have been carefully engineered have Ce values between 0.7 and 0.85.

A final note regarding the measurement of SP_h and VP in hoods is worth mentioning. These values should be considered performance criteria, and as such performance of any system should be measured and maintained. Once a hood is installed, it is important to have access points in the connecting ductwork so the industrial hygienist can regularly measure the SP_h and VP to track the hood performance. Performing regular duct traverses at these access points will provide a plethora of information that the hygienist can then use to quickly deduce if a problem exists. For our example problem, if we know the hood should have 2.89 in. w.c. of static pressure but a traverse reveals that only 2.05 in. w.c. is present, then not enough SP_h is present. The upstream effect is that the flowrate is lower, and therefore, the capture velocity is also too low. The end result is that the employee is not as protected as he or she should be. The fact that the hood "is still pulling" provides a false sense of security. Situations such as these can be rectified by the installation of manometers with a digital

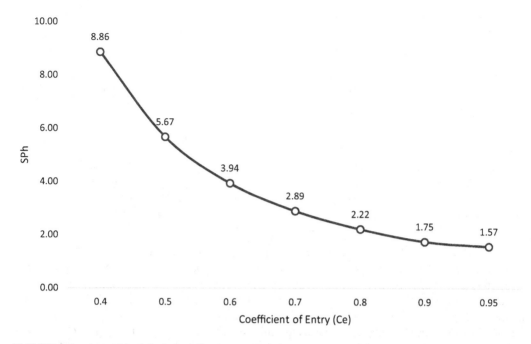

FIGURE 6.4   A graphical depiction showing the SP$_h$ requirement for hoods of various Ce values. For this example, the basis is a flowrate of 2,600 CFM moving through a round 10-inch duct. Note: The Ce values excluded values below 0.4 as it is uncommon to encounter hoods with efficiency ratings so low, although it does happen on occasion.

readout that will alarm if they drop below a certain threshold. While such solutions are common, it is still highly advisable for the hygienist to perform routine evaluations of the hood to ensure performance.

6.3.2   Ductwork

The ductwork of a ventilation system is often an afterthought during the design of an LEV system, considered to simply be the connecting pieces between the hood and the rest of the components. The reality is a far cry from this supposition. Ducts must be given careful consideration as to their design, placement, size, and composition. This section will outline these topics as they relate to the LEV system.

6.3.2.1   Transport Velocity

Arguably, the most important function of ductwork is to safely and effectively transport contaminants from the point of origin (i.e., the hood) to the air-cleaning device. When properly constructed, ducts provide the appropriate transport velocity to perform this critical purpose. Yet to accomplish such a feat, the question that must be answered during the design phase is, "to what transport velocity should the system be designed?". If not designed appropriately, there is a real possibility that the contaminants being conveyed through the system will settle out from the airstream. When this happens, the material can accumulate over time. Accumulated material in ducts can present a variety of hazards, not the least of which is a significant combustibility hazard (assuming the material is combustible), as was discussed in Chapter 5.

The ACGIH has recommended a range of transport velocities based entirely on the physical properties of the materials being handled. These properties focus on the physical state of the material (vapors, gases, dusts, etc.) and the weight/density of the materials. Table 6.2 shows these recommended transport velocity ranges. Most of the examples provided are found in general industries,

TABLE 6.2
ACGIH Recommended Transport Velocities[1]

Contaminant	Recommended Transport Velocity (ft/min)	Examples
Vapors, gases, smoke	Any desired economical velocity (typically 1,000–2,000)	
Fumes	2,000–2,500	Welding fume
Fine, light dust	2,500–3,000	Cotton lint, wood flour
Dry dust/powders	3,000–3,500	Fine rubber dust, cotton dust, soap dust
Average industrial dust	3,500–4,000	Grinding dust, coffee, typical pharmaceutical dust
Heavy dusts	4,000–4,500	Sawdust, lead dust, granulation dust
Heavy, moist dusts	> 4,500	Dust with chips, hygroscopic dusts

and as such are more difficult to directly compare to the contaminants encountered within the pharmaceutical industry. The transport velocities shown in Table 6.2 are frequently cited by practicing industrial hygienists within the pharmaceutical sector. The velocity for vapors and gases is the lowest of the group as those particular phases of matter are assumed to be well-mixed with the airstream, and consequently, there is no risk of the materials settling out. LEV systems conveying only vapors or gases expend far less energy than systems for other contaminants. Concerning pharmaceutical dusts, however, there is a wide range of particle sizes and densities. This often makes it difficult to align desired transport velocities with materials found in traditional industries. Historically, most organizations have leaned toward the 3,500–4,000 ft/min tier of transport velocity, as this range can be broadly used for a wide spectrum of raw materials. The problem, however, occurs if the system designer favors the lower end of the velocity spectrum or if there is inadequate SP being delivered throughout the system and the velocity drops below the recommended range. Moreover, there is always a possibility that a new product or material will be used on a line at some point, potentially requiring higher transport velocities. Furthermore, with an increased focus on process safety within the pharmaceutical industry, modern organizations have tended to mandate higher transport velocities, even for processes utilizing very fine, powdery materials. The typical transport velocity range for dusts within the pharmaceutical sector is between 4,000 and 4,500 ft/min, and this is the recommended range for most pharmaceutical materials. For specialized materials, such as those which are hygroscopic or overtly large granulation materials, transport velocities in excess of 4,500 ft/min should be used. These should be customized to the process at hand and most likely dedicated to the process itself. For the remainder of this chapter and book, any calculations made will focus on attempting to utilize the 4,000–4,500 ft/min range.

6.3.2.2 Duct Size

In the section on hood design, we placed a great deal of emphasis on defining how many CFM are required to make a hood function as intended. Knowing the required volumetric airflow is essential for a hood to perform at its peak, but it is also critical information for sizing the ductwork. Airflow (CFM, or Q) is directly proportional to the linear air velocity (in feet per minute, or ft/min) times the cross-sectional area of the duct (in squared feet, or ft^2). This is represented mathematically by Equation 6.13:

$$Q \left(\frac{ft^3}{min} \right) = V \left(\frac{ft}{min} \right) \times A \left(ft^2 \right) \tag{6.13}$$

This simple but powerful equation contains a wealth of information that is indispensable to the LEV design team; that is, if we know Q (the anticipated CFM needed for the hood to function) and the desired transport velocity (V), we can solve to identify the required area (A) to properly size the duct. All too often, especially on retrofit projects, hoods are connected to existing ductwork via the most convenient or readily available duct sizes without considering transport velocity.

The example from the previous section can demonstrate the utility of Equation 6.13 during the design phase. We had previously tabulated that approximately 2,600 CFM would be required for the flanged slotted hood in our hypothetical scenario. The desired transport velocity is at least 4,500 ft/min. To achieve this velocity, the required duct size needs to be:

$$A = \frac{Q}{V} = \frac{2,600 \text{ CFM}}{4,500 \text{ ft/min}} = 0.578 \text{ ft}^2 \approx 10 \text{ inches diameter} \qquad (6.14)$$

From Equation 6.14, we can see that to achieve the desired transport velocity, the duct must be 0.578 ft^2 in area. For a round duct, this works out to be 10.2 inches in diameter (we will leave the units conversion from ft^2 to inches in diameter to the reader). However, 10.2 inches is not a standard duct size, so the design team would most likely opt for a 10-inch duct, the closest fitting sized duct available. Additionally, by choosing a slightly smaller duct size, the resultant transport velocity is actually higher than the minimum requirement of 4,500 ft/min (the velocity will be approximately 4,700 ft/min).

The overall effect of duct size on transport velocity is significant. Figure 6.5 shows how the transport velocity would change as 2,600 CFM would move through ducts of varying diameters. The velocity drops off considerably as the ducts become progressively larger but eventually begins to plateau to a minimum. The rapid decline in transport velocity as duct size increases is often over-looked during the design phase and is perhaps the greatest contributor to material settling out into the ductwork. A keen observer may note a distinct relationship along the curve in Figure 6.5: as duct size doubles in diameter, the velocity is reduced by a factor of 4. This is highlighted by the velocities at 5-inch (19,066 ft/min) and 10-inch (4,766 ft/min) diameters.

A commonly encountered situation in evaluating older LEV systems is to observe seemingly random and altogether unnecessary ductwork expansions along the system. Situations such as these occur more frequently in quick retrofit projects in which hoods are added to existing lines. When such expansions occur without any additional incoming air, the transport velocity will drop. And as Figure 6.5 demonstrated, the drop in velocity can be significant. Rogue duct expansions are a significant contributor to particulate material settling out of an airstream.

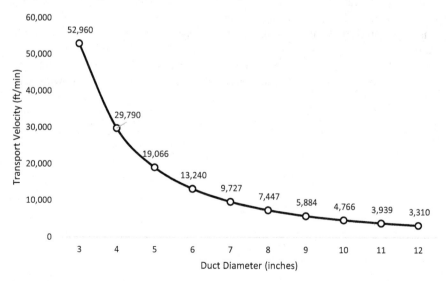

FIGURE 6.5 The change in transport velocity (V) by moving 2,600 CFM through different diameter ducts.

But that is not to say that ductwork cannot expand in route to the fan; on the contrary, expansions are permitted and even required at locations where multiple lines come together, called junctions. When two different airstreams join within a common LEV duct, the resultant volumetric airflow is approximately the sum of the two individual airflows (this is not strictly true, as there are more complicated variables to consider, but this approximate relationship will be sufficient for this discussion). In this manner, if two duct lines have different volumetric airflows, Q_1 and Q_2, when they combine at a junction the result (Q_3) would be:

$$Q_3 = Q_1 + Q_2 \tag{6.15}$$

With a new higher volumetric flow rate, the transport velocity will also increase. The duct should be resized to achieve the desired transport velocity. Figure 6.6 shows an example of how this is achieved. Let's suppose our duct (A) is connected to our previously installed hood which is conveying 2,600 CFM through the 10-inch duct. Along the system, another hood is supplying 650 CFM through a 7-inch duct. They come together into a single duct run at a junction, and the new volumetric flow rate through the duct is 3,250 CFM. If the main duct size (10 inches) is maintained, with the new flow rate, the transport velocity would be almost 6,000 ft/min. While in general faster transport velocities are preferred to convey heavier and denser particulates, this velocity would require more velocity pressure, which in turn requires more SP. This situation robs all upstream operations of that needed SP, making less available for hoods. In addition, unnecessarily high velocities can cause damage to ductwork through abrasion and friction. Where the two lines come together the duct should gradually expand to the desired size in order to achieve the required velocity. In this case, the duct should expand to 11.5 inches in diameter, which would probably be a custom-sized piece of ductwork. But as we have stated before, it is better to invest a bit more money up front to make sure components are designed correctly rather than attempt an expensive fix later.

6.3.2.3   Duct Design

For most LEV systems in existence, the fan which powers the system is not placed close to the hoods. This is usually because fans are large and loud. So for any LEV system, the ductwork is the critical component that links the hoods to the air cleaner and fan. A frequently understated component of the LEV system is how that linkage is accomplished. For older systems in which SP was not given much consideration, ductwork was installed in any which way so long as the hood and the fan were connected. The end result was often a myriad of dips, elbows, bends, expansions, and contractions. Unfortunately, such additions can rob the system of SP that is needed for everything to function properly.

As air passes through segments of ducts that are not straight, the air must turn and dip to follow the path of the duct. When it does, the system uses additional SP to keep the air flowing at the proper

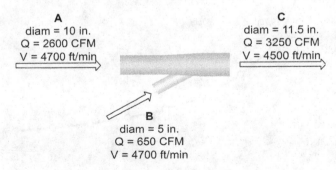

A
diam = 10 in.
Q = 2600 CFM
V = 4700 ft/min

C
diam = 11.5 in.
Q = 3250 CFM
V = 4500 ft/min

B
diam = 5 in.
Q = 650 CFM
V = 4700 ft/min

FIGURE 6.6   The increase in volumetric airflow at a junction requires the duct to be resized to achieve the desired velocity.

velocity. Each type of non-straight ductwork has a unique amount of SP that is required to keep air flowing, and this is a function of the ductwork and its associated loss constant. The loss constants, referred to as f in many texts, are then used to calculate how much static pressure is lost as air flows through these segments. The amount of SP that can be anticipated to be lost as air passes through these areas of ductwork is shown in the relationship in Equation 6.16:

$$SP_{LOSS} = f \times VP \tag{6.16}$$

From the above equation, it can be seen that the amount of SP that will be lost in a specific segment of ductwork is directly correlated to the transport velocity of the air. Intuitively this makes sense: the faster air moves (kinetic energy), the more SP (potential energy) would be required. The loss constant, f, is a unique feature of each piece of ductwork. The more aerodynamic and properly designed a piece of equipment is, the lower the value of f for a segment, and therefore, the lower the SP loss. ACGIH has compiled an extensive array of loss constants for numerous types of duct connections, and these should be consulted whenever ductwork is being installed or designed.

We can demonstrate this relationship by continuing with our hypothetical LEV system. Let's assume that prior to the junction we previously mentioned, the 10-inch ductwork carrying 2,600 CFM has a velocity pressure of 1.42 inches w.c. (from Equation 6.11). At some point, the designer determines that two elbows will need to be added to the duct run in order to convey the system toward the air cleaner (Figure 6.7). One elbow (Elbow A) is a smooth piece of ductwork with a R/D ratio of 2.00 and providing a loss coefficient of 0.13 (loss coefficient values provided from reference texts).[1] In contrast, the second elbow (Elbow B) must be rather tight to make a fairly abrupt turn. The second elbow ends up being a 4-piece segment with an R/D ratio of 0.75, yielding a loss coefficient of 0.33. Applying these values into Equation 6.16, we obtain the following losses in SP through the elbows:

$$ELBOW\ A_{LOSS} = 0.13 \times 1.42\ \text{in. w.c.} = 0.18\ \text{in. SP} \tag{6.17}$$

$$ELBOW\ B_{LOSS} = 0.33 \times 1.42\ \text{in. w.c.} = 0.47\ \text{in. SP} \tag{6.18}$$

These calculated losses in SP are important for designers. Essentially what this means is that as air moves through the elbows, the air slows down and additional SP is required to keep the air moving at the desired velocity. Moreover, the losses are additive; that is, taken together, the design team can expect to lose 0.65 in. w.c. of SP through the elbows alone. Without taking the losses into account, the team will design a system which fails to deliver the needed amount of SP to the hood. If no other losses are present downstream of Elbow B, then it is reasonable to say that the actual value of SP_h that is delivered would be 0.65 in. w.c. less than anticipated, which overall equates to a flow rate of 2,280 CFM, which is 12% below the required value. As it can be seen, failing to properly design the duct system to account for size and SP losses will have a significant effect on the performance of the system (we will see later that a common "solution" to overcome these losses is to increase the fan speed).

FIGURE 6.7 A simplified schematic showing the two elbows, A and B, for our example problem.

It is important to note that every non-straight section of ductwork will have some sort of loss factor associated with it. Each one of these segments must be taken into account and have the approximate loss in SP calculated so that the proper amount of SP can be generated at the fan. Junctions, expansions, contractions, turns, and elbows all contribute to SP loss. Additionally, duct friction also contributes to SP loss within the system. In this vein, the longer the duct run is, the more SP will be consumed as a result of friction. To minimize these effects, modern LEV system designers in the pharmaceutical sector aim to make their systems as short as possible with the fewest turns as possible.

The entire method described in this section is commonly referred to as the ACGIH "balance by design" method. It is now the most widely used method for designing LEV systems and sizing ductwork and has largely replaced older systems which use blast gates to control SP and air velocity. For even simple LEV systems, keeping track of losses from hoods, turns, and other sources can quickly become a difficult and overwhelming task. Fortunately, the ACGIH has a spreadsheet which enables designers to accurately keep track of the losses for each segment of ductwork and to properly size the duct to the desired velocity. The method is also exquisitely detailed in reference 1. The reader is highly encouraged to read more about the "balance by design" method from the ACGIH.

6.3.3   Air-Cleaning Devices

Air-cleaning devices are those pieces of equipment whose job is to remove the contaminants in the LEV air stream. From an environmental and ethical perspective, it is prudent to ensure the air-cleaning device is chosen correctly for the type of contaminant being conveyed. Knowing the type of contaminant present (as well as the anticipated concentration) is vital to choosing the correct type of air-cleaning device. There are numerous types and designs of air-cleaning devices, and entire careers can be made on just the design and installation of these equipment. Air-cleaning devices can be separated into two general categories based on the type of air contaminant present: solid particulates or gases/vapors.

Cleaners for particulates are a common sight not just within the pharmaceutical industry, but in general industry as well. Dust and particulate emissions must be controlled prior to atmospheric discharge. There are four primary types of dust collectors: fabric collectors, centrifugal collectors (cyclones), electrostatic precipitators, and wet collectors. For the sake of brevity, we will focus upon fabric collectors and cyclones as these are some of the more frequently encountered types of dust collectors within the pharmaceutical sector.

Fabric collectors comprise a large and diverse array of dust collectors. These types of devices work by several methods of contaminant removal, including sifting, interception, and impingement of the contaminant onto the filter media. Typically, contaminated air is transported through the ductwork of the LEV system and is passed through and over the various filter media within the dust collector. The efficiency of the filter media is largely dependent upon the material of the filter itself. Importantly, fabric filters are frequently compared by a "permeability" factor, which is defined as the number of CFM that passes through a single square foot of material with a pressure drop of 0.5 in. w.c. Such ratings are vital when assessing a dust collector since air-cleaning devices represent a physical impediment to the airflow from the fan to the hood. There will always be more SP on the "fan side" of the system than on the "hood side" of the system. Knowing how much SP will be lost, or the "pressure drop", as air moves through the air-cleaning device is paramount to designing the LEV system. In this regard, for a fabric dust collector if the permeability is known along with the total surface area and the approximate number of CFM passing through it, the design team can reasonably calculate the anticipated pressure drop and adjust the downstream components accordingly. Yet again, this is an example of knowing the details of the design and how it can impact the entire system. If the pressure drop of the air-cleaning device is not taken into account, the system will experience a significant shortcoming of SP upstream of the device, thus rendering the hood ineffective.

Another aspect of fabric dust collectors is the efficiency of the filters themselves. Many users are accustomed to minimum efficiency rating values (MERV) of filter media but may not associate the values to the efficiency of particulate removal. The efficiency of different filters is dependent on the capability of the filter to remove particulates of specific size ranges. Table 6.3 shows the various MERV ratings and their rated efficiencies for removing particulates. It is important to note the particle size ranges in Table 6.3, as they essentially encompass the thoracic range (10.0 μm), respirable range (4.0 μm), and sub-respirable range (<4.0 μm). As particle size decreases, they are more easily able to penetrate deeper into the lungs, thereby potentially causing more harm. Having filter media that is better able to remove smaller particle sizes is a significant advantage and frequently a requirement within a control banding scheme. Higher-tiered substances (such as OEB 3 or greater) often necessitate the use of HEPA filters to capture as much of the material as possible prior to atmospheric discharge. Yet it is possible that lower-tiered materials, such as benign excipients, may only require MERV 14 filters. The actual requirement is entirely dictated by the organization and its associated risk acceptance criteria, so long as they allow them to meet regulatory emission requirements. When designing a LEV system with a dust collector system, understanding the material and the MERV ratings of the filters is an important risk reduction step.

There is a common phenomenon regarding filter efficiencies that is worth mentioning. A common misconception is that brand-new filters offer the greatest particle capture efficiency. Unfortunately, this is not the case. On a microscopic level, filters have gaps between the fibers. These gaps are what allow air to pass through with resistance, but they also offer smaller particles a greater chance to bypass capture as well. Over time, the filters accumulate material and "plug" the gaps, essentially increasing the capture surface area of the filter. By doing so, the capture efficiency of the filters goes up. Yet this is a delicate balance, because if the filter is never changed, it will completely clog the system and cause damage to the motor and fan. Many dust collectors are outfitted with an air "pulse" cleaning mechanism that knocks some material off of the filter and into the collection vessel, thus prolonging the filter and maintaining maximum collection efficiency. When new hoods are

TABLE 6.3
MERV Ratings and Their Filtration Efficiencies

MERV Rating	0.3–1.0 μm (%)	1.0–3.0 μm (%)	3.0–10.0 μm (%)	Examples
1	<20	<20	<20	Fiberglass and
2	<20	<20	<20	Aluminum Filters
3	<20	<20	<20	
4	<20	<20	<20	
5	<20	<20	20–34	Disposable Filters
6	<20	<20	35–49	
7	<20	<20	50–6%	
8	<20	<20	70–85	
9	<20	>50	≥85	Household AHU Filters
10	<20	50–64	≥85	
11	<20	65–79	≥85	
12	<20	80–90	≥90	Commercial Filters
13	>75	≥90	≥90	
14	75–84	≥90	≥90	
15	85–94	≥95	≥90	
16	≥95	≥95	≥90	
17	99.97	≥99	≥99	HEPA and ULPA
18	99.997	≥99	≥99	
19	99.9997	≥99	≥99	
20	99.99997	≥99	≥99	

installed or if a surrogate powder test is being conducted on an existing hood, it is highly recommended that filters associated with these equipment not be changed for this very reason, as new filters have a greater leak rate and can potentially give false negative tests (we will discuss surrogate powder tests in more detail in the next chapter).

Even though there are devices installed that prolong the life of fabric filter cartridges or media, they do eventually wear out and must be changed. This poses a problem for the individual who must perform the task, as the filters are covered with whatever material is handled in the hood(s) served by the LEV system. In the pharmaceutical industry, this could very easily include API material and potentially even potent compounds (HPAPI). In such cases, there is a risk to the maintenance staff performing the filter change. Modern fabric filter manufacturers have incorporated safe-change systems to account for such scenarios. Often referred to as "bag-in-bag-out" (BIBO) systems, these specially designed filters and their associated housing have built in bags that cover the filters during the change out. The spent, dirty filters are completely contained, keeping the operator safe from exposure and allowing a new filter to be inserted with ease. BIBO filters are standard issue for certain pieces of equipment but must be requested in others. Because they are often specially designed, BIBO filters are significantly more expensive than traditional filters. However, an appropriate risk assessment and control banding scheme will account for this aspect, and the design team should take it into account during the design phase. As before, it is better to spend a little more money up front to design a system properly rather than conduct a retrofit later.

In addition to fabric filters, cyclones are another common form of dust collector within the pharmaceutical industry (and general industry, for that matter). Cyclones operate by removing large particles from an airstream via centrifugal force (see Figure 6.8). Air containing contaminants is swirled through the cyclone, and the force of the air changing direction forces larger particles to eject themselves out of the air stream, colliding against the cyclone, and eventually settling out into a collection vessel beneath. For LEV systems which serve large quantities of materials, such as FIBC unloading or large-scale granulation processes, cyclones are often paired with fabric dust collectors. The reason for the pairing is twofold: First, the astute reader may see Table 6.3 and notice that fabric filters are only rated for particle sizes up to 10.0 μm in diameter. This is because larger particles will quickly overwhelm the filter media, clogging them and rendering them ineffective. The end result is that filters will have to be changed out much more frequently, leading to additional costs in not only filter replacements but the increased down time required for the filter changes. Second, the notion of "larger particles" as it pertains to industrial hygiene typically encompasses

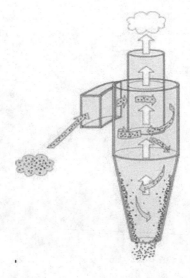

FIGURE 6.8   Rendering of how a cyclone purifies out large particles from an air stream.

those up to 100 μm in diameter (the inhalable fraction of dust). Anything larger is not typically deemed a health concern as they are not easily inhaled. However, as we mentioned in Chapter 5, the risk reduction team must take other hazards into account, and the definition of "dust" when it comes to combustible dust includes sizes up to 500 μm. This is yet another situation wherein knowing the particle size distribution of the materials being handled impacts the system design. If a significant amount of material exceeds 10.0 μm, a cyclone is a prudent first pass means of removing material from the airstream prior to a dust collector. Additionally, the materials collected by a cyclone are large and can accumulate mass rather quickly. So much so, in fact, that it is potentially possible to recycle the acquired material from the cyclone back into circulation, thereby reducing material losses and increasing efficiency (assuming the material continues to meet the criteria from Quality).

Cleaning devices for gases and vapors vary considerably from those for particulates. Whereas dust collectors rely on filter media to gather and remove particulate contaminants, gases and vapors simply pass right through filter media. Consequently, they must be removed through other means. Scrubbers are the most common means of removing vapor contaminants (Figure 6.9). There are numerous types and designs of scrubbers available on the market, but we will focus upon wet scrubbers. These devices operate by moving contaminant-laden air through a column that is densely packed with media. The media has a large amount of surface area across which the dirty air can move and interact. In addition, the column typically has a spray feature which keeps the media constantly moist. Depending on the contaminant in the air, the liquid spray could also be a solution that neutralizes the airborne material or in some other fashion makes increases the cleaning capacity of the column. The dirty air moves through the liquid-laden media and the contaminants are removed, moving ever downward toward a collection tank. The tank can be recirculated to enhance efficiency. The cleaned air moves out through the top of the tower and through the remainder of the LEV system, toward the fan.

Just as with fabric collectors, wet scrubbers also experience a pressure drop. However, the drop across a wet scrubber is often in the range of 1.0 in. w.c. to 3.0 in. w.c. Furthermore, the velocity of the air through the column is significantly slower than through fabric collectors, often in the range of 200–600 ft/min. The slower velocities ensure maximum contact time between the air stream contaminants and the cleaning media. Another aspect for the practicing hygienist to keep in mind

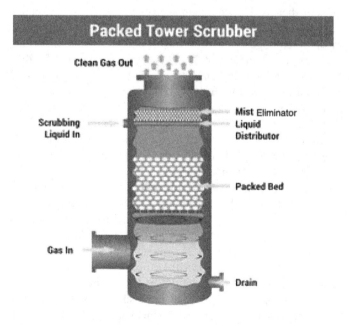

FIGURE 6.9 Diagram of a wet packed tower scrubber system.

is the additional water that is required for the wet scrubber. Such systems often require a volume of 5–10 gallons per minute per 1,000 CFM that flows through the scrubber. While many simple LEV systems for gases and vapors may not utilize a significant amount of air, large systems and multi-component systems can easily add the amount of CFM. In such cases, one must be sure to tabulate not only the additional cost of water (and any other cleaning agents that consist of the solution) but also the cost of disposal for the new aqueous waste stream which contains the removed contaminants.

There are numerous other examples and variations of both fabric collectors and scrubbers. As we previously mentioned, entire careers can be made on designing air-cleaning devices which are suited for a particular industry, process, or material. Adequately cleaning the air stream prior to exhausting is of paramount importance for employees, the environment, and the surrounding community. Ensuring the most appropriate air-cleaning device is chosen is important, but equally important is knowing how much pressure drop will be associated with the device upon installation. These drops must be accounted for during the overall design so that the correct amount of SP is generated and delivered to the system upstream of the fan.

6.3.4   Fans and Motors

Fans are the heart of the LEV system. They are the critical component that drives the entire system, ensuring that contaminants are captured and transported effectively. The sole purpose of fans is to generate enough SP within the system which will overcome all the losses we have already mentioned in this chapter (hood entry losses, elbow losses, expansion losses, sizing losses, friction losses, air-cleaning device losses, etc.). From the generated SP, the fan must also be able to handle the anticipated volume of air at a given rotation rate (RPM). Just as with air-cleaning devices, entire careers can be made surrounding fans and their accompanying electrical motor requirements. There are numerous design types when it comes to fans, and it is worthwhile for the industrial hygienist to be familiar with these types and some of their advantages and disadvantages.

6.3.4.1   Types of Fans

Fans can be broadly categorized into two major categories: axial and centrifugal. Each of these categories can also be further subdivided into more specific variants of the parent categories. Axial fans come in three general designs – propeller fans, tubeaxial fans, and vaneaxial fans. Propeller fans are commonly observed in general industry areas and are used to move large quantities of air without generating a substantial amount of SP. They are typically used for dilution ventilation or area cooling. An example of a propeller fan is the classic free-standing rotating fan. Tubeaxial and vaneaxial fans are similar in concept to the propeller fan but are more robust in construction. Both also move large quantities of air but do so while generating slightly more SP than the propeller variant (between 2 in. w.c. and 8 in. w.c. of SP). For a significant number of industrial operations, particularly those in the pharmaceutical industry, axial fans are insufficient for the desired purpose. Since fans in LEV systems are intended to create SP to overcome losses associated with the system, axial fans are usually not the fan of choice.

Centrifugal fans can also be subdivided into three commonly observed variants – forward curved fans, radial fans, and backward inclined fans. All of the centrifugal fans produce significant amounts of SP, and as a consequence are the fans of choice for LEV systems. Forward curved fans have an advantage of occupying a small footprint and producing less noise than its counterparts, but it is generally not recommended for dusty environments or the potential for dusty environments due to accumulation on the blades. This reduces the efficacy of the fan and increases wear and tear, requiring more maintenance over the lifetime of the fan. For many LEV systems in the pharmaceutical industry, since dust is a potential issue, radial and backward inclined fans are more frequently chosen, with radial fans being selected for environments with copious amounts of dust.

6.3.4.2 Selecting a Fan

Once the type of fan for the system is known, the design team must then select a specific fan that achieves the intended purpose. It is a common misconception that when selecting a fan for a system one only needs to know if it can move the necessary amount of CFM. This is similar to the notion for hoods that if the required CFM is provided, the hood is fully functional and the risk to the employee is reduced or mitigated. However, we have shown that for hoods there is far more involved for hood design than simply knowing the CFM; likewise, for fan selection, there are more details that are required before choosing the correct fan.

In the section on Ductwork, we had mentioned that designing the LEV system to account for all SP losses is the ACGIH balance by design method. This method, when used correctly, sums up all the SP losses along the system and provides the SP generated at the fan inlet (termed SP_{IN}) and also the SP on the exhaust side of the fan (termed SP_{OUT}). In addition, one can tabulate the velocity of the incoming air to the fan, and from this, the velocity pressure (VP) can be calculated (VP_{IN}). These values are used to evaluate the total amount of static pressure needed to overcome all losses within the system and move air as intended. This term is referred to as system static pressure (SSP) and is evaluated according to Equation 6.18:

$$SSP = SP_{OUT} - SP_{IN} - VP_{IN} \tag{6.18}$$

With the SSP known, the design team can then turn to any commercial fan supplier to find an appropriately sized fan along with the necessary electrical requirements (i.e. motor size). Modern fan suppliers have large tables of data which correlate SP requirements to the desired CFM for the system. These tables enable a much quicker and efficient fan selection for a given system. For example, let's suppose our hypothetical flanged slotted hood system has a total SSP requirement of 5.5 in.w.c. (which would be determined after the system was designed in its entirety). We have already determined that the system will convey 2,600 CFM, so the next task is to identify a fan capable of producing this flow rate with an accompanying 5.5 in.w.c. An example fan selection table is provided in Table 6.4. To find the best suited fan, the design team needs to identify the most appropriate CFM rating (left hand column in Table 6.4) and then move horizontally across the table to the desired SP rating. The resulting intersection will provide the necessary speed for the selected motor (in RPM) and the necessary motor size (brake horsepower, or BHP) to power the fan. In our

TABLE 6.4

An Example Fan Selection Chart

CFM	OV	2″ SP		4″ SP		6″ SP		8″ SP	
		RPM	BHP	RPM	BHP	RPM	BHP	RPM	BHP
1,000	1,580	1,227	0.59	1,641	1.13	1,964	1.70	2,245	2.34
1,325	2,093	1,312	0.87	1,715	1.55	2,031	2.24	2,306	2.99
1,650	2,607	1,409	1.22	1,793	2.05	2,108	2.90	2,367	3.74
1,975	3,120	1,515	1.64	1,885	2.66	2,183	3.65	2,444	4.65
2,300	3,633	1,631	2.16	1,984	3.37	2,274	4.53	2,532	5.71
2,625	4,147	1,754	2.79	2,089	4.18	2,371	5.54	2,614	6.84
2,950	4,660	1,881	3.55	2,201	5.13	2,467	6.64	2,713	8.18
3,275	5,174	2,011	4.44	2,317	6.20	2,576	7.91	2,813	9.64

Note: CFM is the desired CFM moving through the system which the fan must be able to handle; OV is output velocity, or the velocity of the air as it leaves the fan; RPM is the speed of the fan in revolutions per minute; BHP is brake horsepower, or the size of the motor required to power the fan.

hypothetical case, the most prudent choice would be the 2626 CFM option paired with the 6" SP, yielding the requirement of 2371 RPM with a 5.54 HP motor.

It is worth noting that designers often build in safety factors to account for "unanticipated" occurrences. For instance, miscalculations on SP requirements or last-minute changes to branches can require additional capacity. For these reasons, and others, fans with a higher capacity and a larger motor are often chosen for a given system. Such safety factors are often necessary in real-world scenarios.

6.3.4.3 Fan Curves, System Curves, and System Operating Point (SOP)

As we have mentioned, there are numerous types and variations of fans, each with their own unique performance characteristics. The data found in the lookup tables from manufacturers is specific to the performance of that fan at specific operating conditions. All fans are tested according to specific protocols, such as those found in ANSI/AMCA 210.[2] For a given fan and test, ideal conditions are established such as aerodynamic hoods with minimal losses, long straight lengths of duct leading into and away from the fan, etc. Under each test condition, the RPM of the fan is held constant, and the hood is initially blocked, maximizing the SP generated by the fan. Incrementally, the inlet is unblocked and the SP is measured along with the volumetric airflow. These steps are repeated until the inlet is completely open and the maximum airflow is achieved. The test can be repeated by altering the fan speed prior to restarting the test. If the results from each fan test were displayed graphically, the result would be what is called a fan curve. Each fan curve is unique to a given fan at a specified RPM and duct size. They provide visual cues as to how the fan performs over a range of SP and airflows, with some having sharper drop offs than others. The fan curve is a critical piece of information that should be requested by the design team and kept on file. It is important to realize that fan curves are applicable only at the RPM specified and that they were generated under ideal conditions. Most real-world installation scenarios are far from ideal, so fan performance will undoubtedly be different once installed (more on this later). A hypothetical fan curve is provided as an example in Figure 6.10.

Separately, each LEV system has a unique characteristic that relates the system static pressure (SSP) to the total volumetric airflow (CFM) served by the system. In essence, this relationship is a

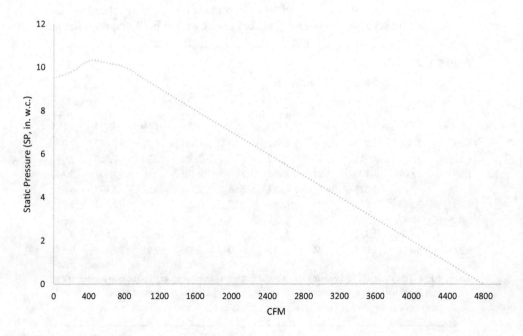

FIGURE 6.10   An example of a hypothetical fan curve.

measure of how well the system overall uses SP to facilitate and convey the transport of air in CFM. This concept is very similar to the coefficient of entry (Ce) we discussed earlier as it relates to hoods and their efficiency to convert SP into VP. Mathematically, the relationship between SSP and CFM is shown in Equation 6.19.

$$SSP = K \times Q^2 \tag{6.19}$$

System static pressure is equal to the square of the total volumetric airflow (Q) multiplied by a constant, K. The constant is unique to each LEV system and is a function of the overall resistance in the system, construction, efficiency, etc. From Equation 6.19, it is possible to calculate the amount of SSP needed for a desired volumetric flow rate. Once an LEV system has been fully designed (hoods, ductwork, air-cleaning devices, and exhaust stacks), SSP and Q can be calculated and therefore inserted into Equation 6.19 to solve for the system coefficient K. Once all variables are known, they can be graphed to display what is known as the system curve. System curves are parabolic relationships which visually showcase how these two variables relate to each other within the system itself. It is important to note that while a system curve is unique to each LEV system, any changes to the design, layout, or resistance within the system will alter the system curve, generating a new one. This is the situation when systems are designed with dampers or blast gates. By opening or closing dampers, the pressure resistance at the particular junction changes, either increasing or decreasing, and by extension, the overall airflow either increases or decreases. An example system curve is shown in Figure 6.11.

So why is knowing the fan curve and system curve important? And how do they relate to each other? While they are independent of one another, they do affect one another. If the fan curve and the system curve are graphed together, they will (usually) cross at some specific point. That point of intersection is termed the system operating point (SOP). The SOP is the point where the systems work the most efficiently: the fan supplies enough SSP for the LEV system to overcome the losses and resistance throughout the system, while the system provides the CFM needed to balance out the fan and make it operate smoothly. Figure 6.12 is a graphical overlay showing how Figures 6.10

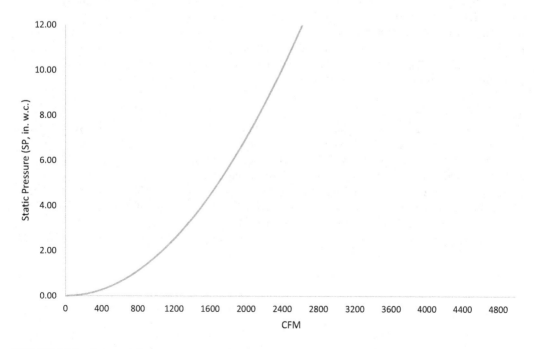

FIGURE 6.11   An example system curve.

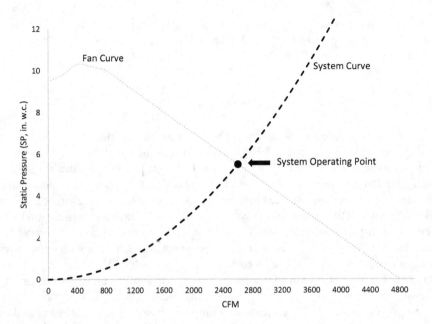

FIGURE 6.12   An overlay of the fan curve and system curve, showing the system operating point (SOP) at the intersection of the two curves.

and 6.11 overlap. In this example, the two curves intersect at 2600 CFM and 5.5 in. w.c., the desired flowrates and SSP requirements we had stated earlier that were required to make the designed LEV system nominally operate.

When the two curves intersect at the desired SSP and CFM, the system will be well-balanced and operate as intended. But it is possible to have these two curves intersect at a sub-optimal location. For instance, if the cheapest fan is purchased without knowing how it truly performs, the system curve will intersect at a completely different performance, giving rise to an inefficient system that may not deliver the desired airflow. Furthermore, if the system curve is wrong (due to more or less resistance present in the system), then system curve will alter, giving rise to a similar scenario. In both scenarios, the results are not ideal and a new fan must be installed or the system redesigned. This is why it is so important to properly calculate the losses and resistance within a system and to select the correct fan.

6.3.4.4   The Fan Laws

We mentioned earlier that fans are tested and rated under ideal laboratory conditions, yet real-world conditions are almost guaranteed to not be ideal. All too often, if an LEV system is not performing as intended with not enough airflow at the hood, the solution is to "increase the fan speed". In principle, this solution may seem completely justified: the faster a fan spins, the more air it will move. While this is conceptually correct, there are other downstream effects which can have significant consequences on doing so. Specifically, increasing the fan speed has an effect on flow rate (CFM), static pressure (SP), and the energy requirement to the fan motor (horsepower, HP). These are related by what are known as the Fan Laws. These relationships are shown in Equations 6.20–6.22.

$$CFM_2 = CFM_1 \times \left(\frac{RPM_2}{RPM_1} \right) \tag{6.20}$$

$$SP_2 = SP_1 \times \left(\frac{RPM_2}{RPM_1} \right)^2 \tag{6.21}$$

$$HP_2 = HP_1 \times \left(\frac{RPM_2}{RPM_1} \right)^3 \qquad (6.22)$$

Since fans are tested with their fan speed (RPM) held constant to generate the fan curve, increasing or decreasing the fan speed essentially gives rise to a brand-new fan curve. The volumetric flow rate is directly proportional the change in fan speed; that is, a 10% increase in RPM yields a 10% increase in CFM. However, the change in SP is related to the square of the fan speed change. Consequently, increasing the fan speed by the previously stated 10% yields a 21% increase in SP. But the most drastic change occurs with the fan motor. The change in horsepower is related by the cube of change in fan speed. As such, the 10% increase in fan speed yields a 33% increase in horsepower requirement to power the fan. The utility of the Fan Laws allows the practicing industrial hygienist to quickly estimate how changing the fan speed will impact the system and judge how it may or may not be a viable solution to a given problem.

This can be demonstrated with an example. Let's assume that our LEV system with the flanged slotted hood is installed as designed with the chosen fan and motor (2371 RPM and 5.5 HP). After installation, the hood is tested and is shown to be pulling only 2,000 CFM and having a SP_h of 1.5 in. w.c. Since the utilities are below what we want, the supervisor to the area requests that the fan speed be increased. Using the Fan Laws, we can see that an increase from 2,000 CFM (measured) to 2,600 CFM (desired) is an increase of 30%, or a factor of 1.3 (this is the value of the ratio RPM_2/RPM_1). When applied to the SP requirement, we get an increase of 69%, bringing the SPh value from 1.5 in. w.c. to 2.54 in. wc. This value is below the previously calculated requirement of 2.89 in. w.c. for the hood, but overall it still provides a reasonable transport velocity (4,400 ft/min) through a 10-inch duct. However, the biggest change is to the energy requirement. A fan speed increase of 30% yields a horsepower increase of almost 120% ($1.3^3 = 2.197$). When applied to our original motor of 5.5 HP, the new energy requirement to power the fan is 12.1 HP. This outcome precludes the possibility of increasing the fan speed enough to achieve the target airflow. While the notion of "increasing the fan speed" seems to make sense, by doing so the motor would burn out. It makes more sense to either resize the fan, install a larger motor, or both.

Yet even if the motor could withstand the increased electrical demand placed upon it, we again arrive at an intersection wherein sustainability plays a key factor. Most major pharmaceutical organizations have initiated programs to try and reduce the amount of electricity utilized at their sites. Having an issue in which the solution is to increase power consumption by over 100% does not assist with the overarching goal. The easy solution may be to increase the fan speed as much as possible, true, but long-term this is not only harmful to the sustainability goals, but it also places additional strain upon the fan and the motor. There are countless models of fans that can generate the required SP requirements while using a fraction of the power as a result of a different design. It may cost a bit more to have a properly sized fan and motor, but the long-term result far outweighs the subsequent changes made by altering fan speeds and changing fan curves and system operating points.

The methodology presented here for fan selection is a very simplified process. In reality, there are numerous details that must also be considered, such as the layout of the fan, the aerodynamic properties of the wheel, decisions regarding heat capacities, material of construction, the number and types of blades, compatibilities with contaminants, and even noise considerations. The use of a table to pick a fan can easily be misread, leading to an incorrect fan to be selected by those inexperienced in the process. Most fan companies have freely available software that allows the designer to effectively narrow down the selection. It is highly advisable to take advantage of such tools whenever the fan selection process is undertaken. Perhaps even more, though, is the recommendation to seek out an engineer experienced in LEV design. The aforementioned fan companies also have knowledgeable staff that can aid in the selection process. Finally, whenever a fan has been chosen, it is critical to keep on file the model chosen for maintenance purposes. All fans wear down over time, so knowing how to replace one is critical. Furthermore, having a copy of the fan curve enables the maintenance group and the industrial hygienist to quickly measure and troubleshoot for problems.

6.3.4.5   System Effect Losses

A frequently encountered phenomenon with LEV systems is the system effect loss. In essence, this is a change in the fan's performance as a result of the design of the LEV system. Up to this point, we have treated the system (ductwork, hood, and air cleaner) as a separate entity from the fan, with independent performance curves which need to align for optimal performance. In reality, the design of the system directly impacts the performance of the fan. When system effect losses occur, the result is that the real-world fan performance curve changes, making it less effective.

The cause of a system effect is uneven airflow at either the inlet or the outlet of the fan. As we mentioned before, fans are tested under ideal laboratory conditions wherein long, straight lengths of duct lead into and out of the fan. These straight lengths of duct allow the air to become evenly distributed and load upon the fan evenly in the middle. This uniform distribution upon the fan lowers the impact on the blades and the various parts of the fan, creating less resistance and thereby generating the maximum amount of SP possible for the fan. When systems are designed such that bends, elbows, or 90° turns are installed immediately prior to air entering the fan, air flow becomes extremely turbulent and the airflow spirals as a result of the elbow construction. If the spiraling airflow is in a direction opposite the spin of the fan, it must now work harder to change this trajectory, requiring more SP to overcome this effect. The usage of SP to correct this airflow issue results in less SP available upstream to the hood.

System effect losses are also created from the discharge side of the fan, or the exhaust. As air leaves a fan, there are zones of varying velocity. These non-uniformly distributed airflows require long straight runs of ductwork to even out. Unfortunately, a common occurrence on exhaust systems is for design teams to install elbows close to the fan in an effort to divert the airflow in a desired direction. The placement of such elbows causes significant disruptions in the airflow, resulting in eddy currents within the elbows and even some zones to travel backward toward the fan. In these situations, the fan must once again work harder to overcome the additional resistance applied to it, lowering the overall SSP. Both sources of system effect losses, resulting from disruptions to the inlet and the outlet, are often a direct result from area and size constraints where the system is being installed. Consequently, system effect losses are one of the most commonly encountered sources of lower performing systems.

So how do system effect losses actually affect the fan curve? As we previously mentioned, fans are tested under ideal conditions with uniform airflow both into and out of the fan. The tests generate the fan curve which is the basis for fan selection. But the inclusion of system effect losses greatly alters the fan performance and generates what is, in effect, an entirely new fan curve based on the system which it is serving. The result is a fan curve which will perform less effectively than anticipated. When overlayed with a system curve, the difference between the fan performance curves becomes apparent. The initially desired SOP cannot be achieved with the adjusted fan curve; rather, a new, underperforming SOP results, providing insufficient CFM and SP for the system. In order to achieve the airflow that was deemed a requirement for the hood to operate, the fan speed would need to be increased (see the Fan Laws). But increasing the fan speed to the needed RPM would result in the operating point being "off the fan curve". This disparity, where the operating point is not set at the SOP and is in fact "off the fan curve", is the system effect loss. To make the fan operate under these conditions places a significant amount of strain upon it, leading to fan cracking or other components wearing out faster than anticipated.

System effect losses can be minimized by appropriate design. Long, straight runs of duct both into and out of the fan will greatly reduce these losses and ensure the fan properly functions for a long time. The exact lengths required vary, but a general rule of thumb is 3–5 duct diameters of straight duct going into the fan after any elbows and out of the fan before any elbows should be sufficient to minimize such losses. In effect, this means that if a 12-inch diameter duct is installed to carry air to a fan, then straight ductwork in the length of 36–60 inches should be a part of the design. For most enterprises, space is a significant constraint and that much square footage may not be available for implementation. This is the source of most, if not all, system effect losses: the lack of available space for appropriate design. As a consequence, most LEV systems are forced to implement the tight turns and elbows connecting directly to the fan that we have stated should be avoided. In such cases, the practicing IH and the design team must be aware that system effect losses will be present, but the extent of them cannot be predicted. If at all possible, efforts should be made to install the proper lengths of ductwork to prevent system effect losses.

6.3.5 EXHAUST STACKS

The final component of any ventilation system is the exhaust stack. Often an afterthought during the design phase, exhaust stacks are frequently given little attention so long as the airflow is directed out of the building. This is a significant misstep, as thought and care should be placed into not only where the exhaust stacks should be placed, but how high they should be, and exit velocities. Such features are important, and the site industrial hygienist should have a good understanding of how these may impact the system and the building itself.

6.3.5.1 Exhaust Stack Placement

The placement of an exhaust stack is one of the first design items that must be addressed for the final leg of the LEV system. A common implementation involves putting the stack close to the fan but placed on the roof and connected by as little ductwork as possible. Such designs are easier to install and typically cost less but may not represent the most appropriate strategy. A common shortcoming regarding exhaust stack placement is the proximity to air intakes for HVAC units. Far too often the exhaust of LEV systems is placed close enough to the intakes to allow for some of the exhaust to be re-entrained into the building. Such occurrences are often the source of building occupant complaints regarding odors or other indoor air quality issues. A commonly used rule of thumb for exhaust stacks is to place them at least 50 ft away from any air intake. Such a placement helps to ensure that exhausted air is kept out of the entrainment path (although this is not an absolute guarantee based on distance alone).

This distance requirement poses a potential design issue for the risk treatment team. It is rather common for designers to run ductwork up to a roofline or penthouse from the process being controlled in a vertical fashion; that is, wherever the process occurs in the building, the design team often simply goes up with the ductwork to the roofline or the penthouse to place the fan, motor, and exhaust. Such a strategy shortens the system and effectively costs less money. Yet by doing so, there is no guarantee that the ultimate placement of the exhaust stack will be a suitable distance from an air intake. It is far better to have a designated location on a roof where virtually all exhaust stacks should be located, or at least where intakes should be avoided.

This begs the question of knowing where to collect any and all exhaust stacks and where to avoid placing air intakes. This is largely dictated by wind patterns, as this is the source of moving exhaust air into the HVAC system. Most geographical locations (i.e., cities) have years of published weather data wherein the average speed and direction of wind are collected and presented collectively as a wind rose. Wind roses display the frequency of direction the wind is blowing from as well as how often that wind blows with a certain velocity (wind roses can be found from various meteorological groups such as the NOAA). This information can prove useful to the LEV design team since it will allow them to place exhaust stacks downwind from air intakes, thereby reducing the probability of entrainment into the HVAC system.

6.3.5.2 Building Effects, Stack Height, and Exit Velocity

Stack placement is but one piece of designing the exhaust component of the LEV system. It is also important to take building effects, stack height, and exit velocity into consideration. The building from which the process is being exhausted has a physical impact on the roof. Perpendicular wind hitting the building invariably causes a myriad of eddy currents, vortices, and downwash effects from the exhaust stacks. These effects work together to create various zones wherein improperly placed exhaust stacks can introduce contaminants which get brought back to the roofline where they can be entrained back into the HVAC system or expose personnel who may be working on the roof.

To ensure that exhaust stacks do not discharge air into such zones, the height of the exhaust stack must be designed appropriately. There is no single rule which stipulates how high an exhaust stack must be, but there are means of estimating how high a stack should extend to escape recirculating areas on a rooftop. These estimations are based on the size of the building and any surrounding obstacles which may interfere with the airflow patterns. Generally speaking, exhaust stacks should be placed on the highest possible point of a rooftop to ensure that a stack extends beyond any recirculating zones; otherwise, stacks can be 10 ft tall or higher, depending on the physical nature of the building. A more detailed discussion on estimating effective stack heights can be found in Reference 1.

In addition to ensuring a proper stack height, an equally important feature is the discharge velocity of the exhaust air. If traveling fast enough, the exhausted plume of air will extend a significant distance into the atmosphere, thereby increasing the "effective stack height". A frequently utilized rule of thumb is that the exhaust stack (height from the roofline) should be 1.5 times the maximum air velocity experienced on the roof. The term "maximum air velocity" is open for interpretation, but from a design perspective, a guideline can be the 95th percentile of air velocities measured in the area. Again, meteorological data will be helpful in this endeavor. For example, if the typical maximum wind velocity experienced in the area is determined to be 18 mph (1,584 ft/min), then the exhaust stack should have a minimum discharge velocity in excess of 2,300 ft/min, although higher velocities may be preferred (see below).

Properly designed discharge velocities also offer a distinct secondary advantage. Historically, many discharge stacks were equipped with "rain caps", small hats that upon the stack to prevent rain and other precipitation from entering the stack. While effective in this endeavor, the rain caps placed a physical barrier in front of the exhausted air stream, thereby completely altering its trajectory, slowing it down, and adding another source of resistance to the exhaust side of the fan. By altering the path of the exhausted air, rain caps force the air stream back onto the roof, causing it to recirculate. On the off chance, the air-cleaning device did not do a sufficient job of cleaning the air stream, anyone who needed to perform work on the roof would be at risk of exposure as a result. It has been effectively demonstrated that exhaust stacks with discharge velocities in excess of 2,600 ft/min are able to prevent rain from entering the stack. If it is not possible to design the discharge velocity to this level, or if the system is not run 24/7, then exhaust stacks can be designed with off-set elbows, or slight curves which collect rain and divert the water away from the fan and the rest of the system. Stated another way, it is no longer an accepted practice to simply place rain caps on discharge stacks; rather, the system should be designed in such a manner that the discharge velocity of the air prevents rain and precipitation from entering, or to install offset elbows in the design.

6.4   SUMMARY

This chapter has taken a look at the five components which comprise an LEV system: the hood, ductwork, air cleaner, fan, and exhaust. Within a proper risk management framework, designing and implementing (or simply updating/retrofitting) an LEV system comprises the risk treatment phase. The overall risk to the employee/operator has been identified and thoroughly documented through quantitative sampling and various statistical analyses of the acquired data. This phase of risk management if undertaken when the risk management team decides that the appropriate course of action is to reduce or mitigate the risk level (i.e., treat the risk) rather than another course of action. Within the hierarchy of controls, engineering controls are the preferred method of risk treatment (assuming the hazard cannot be eliminated or substituted out). For most engineering controls, the LEV system is the heart and soul of the system and the primary means by which contaminants are removed from the workplace. Therefore, it is absolutely vital to understand how these systems work and the appropriate means to design and implement them effectively.

To make matters more complicated, the designer of the system cannot simply purchase the cheapest available option or nonchalantly insert bends and elbows into a system, nor can he or she buy any fan and simply increase the speed, hoping to achieve target performance. A great deal of information is required at the outset of the design and even after installation to ensure the system is performing properly. After all, if the LEV system is not functioning properly, then the risk is not being effectively treated. Industrial hygienists within the pharmaceutical sector have gone to great lengths to understand and characterize all the LEV systems in their facilities to ensure the appropriate level of risk treatment is being implemented. For organizations which may be undertaking efforts to upgrade existing systems or construct new sites with new ventilation requirements, it is highly recommended to bring an industrial hygienist into the conversation. The practicing hygienist can offer significant input on design aspects and help to ensure that the system will deliver on its promised goal.

NOTES

1 ACGIH. *Industrial Ventilation: A Manual of Recommended Practice for Design*, *30*th Ed. 2021.
2 ANSI/AMCA 210–16: Laboratory methods of testing fans for certified aerodynamic performance rating.

7 Risk Treatment
Hoods and Containment

7.1 INTRODUCTION

The last chapter was an overview of the entire local exhaust ventilation (LEV) systems as part of the risk treatment process (Figure 7.1). LEV systems are comprised of five individual components that must together remove contaminants and hazards from the employee's work zone: hoods, ductwork, air-cleaning devices, fans/motors, and exhaust stacks. Each component is critically important to the overall performance of the system. However, hoods are the starting point of the entire system, the part where the contaminants first enter the LEV system and whose design directly impacts downstream efficacy. Hoods are also the LEV system component with which operators directly interact. Their design and performance have a direct correlation to protection of employees more than any other component. Since there are countless variations of processes, hood designs come in just as many unique designs to accommodate the specifics of a given situation.

But hood design and performance are not dictated solely upon how a process is performed. The biggest driving force behind hood performance is the required level of protection for the operator, which in turn is dictated by the OEL or the specific control band in which the material is placed. Such information is typically available from doing the first two steps of the risk assessment process (Hazard ID and Dose-Response Assessment) and is critical during the selection of a hood.

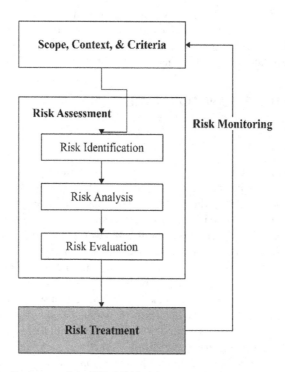

FIGURE 7.1 One of the final steps of the ISO 31000 risk management flow is risk treatment.

DOI: 10.1201/9781003273455-7

Throughout this book, we have occasionally used the term "containment" when discussion a hood's ability to prevent unwanted chemical exposures. Historically, containment referred to completely isolating a process from the surrounding environment. Typically, such isolation was achieved with full enclosures, such as gloveboxes. This was the prevailing line of thought for years. Whenever the recommendation for containment was brought up, the assumption was that gloveboxes would be involved. Since these enclosures are typically used for processes involving HPAPIs, many end users would balk at the idea of containment and instead defer to cheaper or more readily accessible options.

The concept of containment has undergone a paradigm shift in the last few decades. Rather than referring strictly to isolation, containment is now meant as the prevention of material from escaping a work area or from reaching the operator. These concepts, enclosure versus prevention of escape, may seem as though they are the same definition simply worded in a different way, or perhaps we are splitting hairs with the wording. But when put into practice, there are significant differences between the two. Isolation requires the user to work within a box, such as an isolator. However, hoods exhibiting high levels of containment can also be those that prevent the material from entering the breathing zone of the operator while still being "open" to the workspace. There are a variety of hoods which achieve this feat for a variety of processes, as we will see. Such efficient hoods have ushered in new eras of safety as they allow for enhanced protection while making operations more comfortable for operators with ergonomic improvements.

Given that hoods play such an important role in the risk treatment phase of risk management, this chapter will focus on various hood types used in the pharmaceutical industry and the level of containment typically offered by such options. Information regarding their use and utility requirements will be presented as well. Our focus for this chapter will be solely upon hoods and the typical operations performed within them. It should be noted that there are numerous opportunities for containment to be applied in other processes beyond hoods. For example, material transfer in bags or containers, material conveyance, tablet presses, milling operations, granulation processes, and packaging are all aspects of the pharmaceutical chain that require various aspects of containment. To delve into these aspects with detail is beyond the scope of this book. The ISPE put forth a publication, *Good Practice Guide: Containment for Potent Compounds*, which details numerous examples of containment surrounding multiple aspects of potent compounds and for a multitude of pharmaceutical processes.[1] However, we will restrict our discussion to a few select examples of hoods for brevity. Finally, a brief discussion on ways to validate hood performance will be included as well.

7.2 CONTAINMENT OPTIONS FOR HANDLING SMALL QUANTITIES OF MATERIALS

7.2.1 SNORKEL TRUNKS

Once upon a time, snorkel trunks were one of the most commonly utilized hoods in virtually every industry, including the pharmaceutical sector. Sometimes referred to as an "elephant trunk", these hoods are easily identified by their elongated flexible trunk. Snorkel trunks found significant popularity among operators because they can be easily placed in virtually any configuration. Theoretically, as a process or material moved about in the workplace, so too could the operator relocate the trunk. In addition to operators favoring their use, designers and managers also commonly give snorkel trunks a ringing endorsement due to their low cost and ease of installation and replacement. Snorkel trunks are unquestionably among the least expensive options available on the market. This fact makes them popular options for quick retrofit projects in which ventilation is deemed a requirement, but no forethought is used to determine how much airflow is required. Snorkel trunks can be found as implemented hoods for a variety of processes, including sampling, weighing, powder discharging, liquid filling, transfers, and milling.

Despite their apparent endorsement from workers and management alike, snorkel trunks have fallen into disfavor among pharmaceutical industrial hygienists, especially for processes involving powders. The reasons for this are many. First, for operations involving snorkel trunks, the material is being actively released into the workspace and the hope is that the trunk will capture enough of the material before it reaches the operator. But the level of containment offered by snorkel trunks is extremely variable, typically ranging from 1,000 μg/m^3 (or higher) to as low as 200 μg/m^3 for powders. In practice, most operations will be on the higher side of this range. If we were to assign these containment levels to a control band or occupational exposure band (OEB), it would be restricted to OEB 1, or non-hazardous/innocuous materials. Snorkel trunks are typically not permitted for use with materials in higher-tiered bands because there is virtually no effective dust containment provided.

Recall from Chapter 6 wherein we stated that the effectiveness of LEV hoods is dependent upon the distance from the emission source. Snorkel trunks are at the mercy of this phenomenon perhaps more than any other type of hood. They must be placed extremely close to the source, often within mere inches of the operation to be effective. When the trunks are moved, their placement is not typically chosen for their effectiveness but rather for their convenience. This high degree of variability with trunk placement is the primary reason that the level containment is so varied. Yet there are other reasons for their predictably unpredictable performance as well. The flexibility that is often viewed as an advantage by the users works against the effectiveness of the hood. Rigid systems can withstand pressure changes and convert SP to VP effectively with minimal friction losses. Snorkel trunks, however, lose significant amounts of SP with the frequent dips, turns, and even pinches along the trunk length. In addition, since these hoods are often composed of plastic that contains ridges, there is significant friction losses within the trunks. When taken together, snorkel trunks have very low Ce values, often below 0.4. To overcome these losses, a significant amount of SP would need to be present in order to generate the needed capture velocity and CFM required. But since these are often added without conducting pressure calculations, the overall flow rate is often ineffective for powders.

For these reasons (and others), snorkel trunks are being phased out as the go-to option for hood installations. Their overall poor performance for powder handling operations which restricts their use to only non-hazardous materials on small-scale processes severely limits their utility. As we mentioned in previous chapters, APIs are becoming increasingly potent and a hood which cannot provide adequate containment levels for operators is simply not an option in these scenarios. In many pharmaceutical organizations, snorkel trunks are being relegated to specific non-API tasks such as welding (perhaps the most common use of snorkel trunks in any industry) and operations within laboratories, such as controlling exhaust from instruments. For these operations, since the emission source is highly localized and usually generated in a singular direction, snorkel trunks are actually highly effective in these situations.

7.2.2 Downflow Booths

Downflow booths are large semi-open areas with a single open area while the entire back wall and the ceiling are, in effect, the hood and the source of containment (Figure 7.2). The booths can conveniently be custom made to fit virtually any operation, but within the pharmaceutical industry, downflow booths are typically relegated to processes involving sampling, discharging, or weighing powders and solids. The configuration for a downflow booth is rather popular because it provides what most containment devices do not: operational space. Because the "room" is the hood, operators can freely move about, within reason. Typically, operators highly favor downflow booths.

The containment provided by the downflow booth stems from a highly filtered recirculating air current (Figure 7.3). A fan generates substantial airflow at the bottom of the backwall of a downflow booth, thereby creating a downward draft which pulls airborne contaminants down and away from operators. The air passes through a pre-filter and cycles back into the ceiling, at which point it passes through a series of HEPA filters before it is reintroduced back into the hood space. The direction of the air stream is from top to bottom, thereby creating another avenue of "pushing" contaminants down away from the operator's breathing zone and into the air stream toward the capture zone. The hoods typically have the front face of the room open to allow for some new air to enter the space, while the booth diverts some used air into the ductwork and toward the air cleaner. But the majority of air utilized in the downflow booth is recirculated. The recirculation concept is pivotal to making the downflow booth cost effective as it minimizes the amount of makeup air required for the unit.

FIGURE 7.2   An example downflow both. Downflow booth image courtesy of the Dec Group.

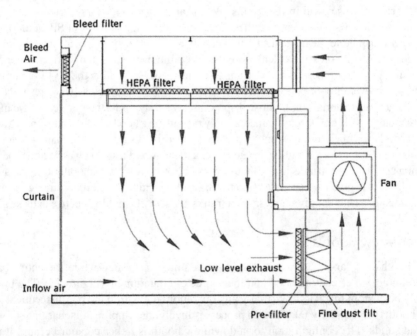

FIGURE 7.3   A diagram showing the airflow patterns and source of containment for a downflow booth.

The containment level offered by downflow booths is highly variable, dependent upon amount of material being handled, particle size, rate of operational speed, and duration of the task. Many downflow booths are used for material handling processes, such as weighing, sampling, or mixing. A good portion of these activities are restricted to smaller scales (such as 50 kg or less), making the possibility of containment much more likely. When used properly and according to design specifications, many downflow booths can provide consistent containment levels on the order of 100 μg/m³, but these claims should be verified by the manufacturer and also tested upon installation to ensure levels can be achieved.

There are some drawbacks to the use of downflow booths. The first is the energy requirement. Fairly powerful motors are needed to move the large volumes of air required for the hood, which can be a burden on energy-conscious organizations. A second drawback is that operator placement within the hood has a major impact upon the efficacy of hood performance. These hoods are designed to be used as close to the hood face (that is, as close to the back of the hood nearest the fan) as possible. This region of the booth has the greatest capture velocity and subsequently offers the most protection to the operator. Experience has shown, however, that operators may sometimes arrange equipment in such a manner that is most convenient, which may not align with the intended design of the hood. In these situations, the containment ability of the hood can be severely compromised. Another drawback is that during the operation, the materials being handled are still potentially being released into the general workspace of the operators. The hope and intent is that the hood generates enough velocity and capture capability to remove the dust and particulate from the area before employees are overexposed. Because materials are released into the general workspace, operators typically still utilize full protective suits, gloves, and possibly even respiratory protection (such as N-95 respirators or half-face respirators). The required use of PPE may allow for somewhat more potent materials to be handled, but this is typically not allowed, and downflow booths are often restricted to OEB 1 or 2 materials. A final drawback of a downflow booth is that proper maintenance is crucial to their long-term effectiveness. The motor and fan which powers the booth must be in prime working condition to generate the necessary airflow. Maintaining regular preventative maintenance on these components ensures that the optimal containment is sustained over the working lifetime of the hood. Likewise, the HEPA filters which are vital to removing contaminants from the recirculated airstream must be changed regularly. Importantly, the new HEPA filters must not only be seated correctly in their housing but must be free of tears or holes. In such situations, contaminants will escape through the punctures and a subsequent loss of containment will be experienced.

7.2.3 Ventilated Balance Enclosures (VBE)

In the modern pharmaceutical organization, VBEs are ubiquitous (Figure 7.4). A VBE, or ventilated balance enclosure, is a modern-day engineering control which utilizes laminar airflow to achieve containment. They have become the workhorse of many small-scale pharmaceutical operations including sampling, weighing, dispensing, and material analysis. Their popularity stems from the excellent level of containment they offer. Depending on the scale and robustness of an operation, VBEs can provide containment often below $1 \, \mu g/m^3$. For many control banding schemes, this level of containment is suitable for substances in the OEB 4 range. Simply put, the level of containment offered by VBEs allows for a much wider range of materials to be handled.

Because VBEs have become so popular, they are commercially available from numerous manufacturers in a variety of off-the-shelf assemblies. Such assemblies can include integrated scales, waste chutes, and cutouts for drums of materials to be brought into the enclosure. Importantly, VBEs have an open face that allows for operator mobility. The enclosures are most often purchased as standalone units for a single role or purpose, but there are many examples of customized VBEs. For instance, it is not uncommon for a VBE to be coupled with another enclosure for a subsequent step. Weighing operations often take place within a VBE, and the material is then transferred to the second enclosure via a connected pass-thru port. In such a scenario, the chemical substance is completely handled within the containment system.

It is noteworthy that for VBEs, the parent container is opened and handled entirely within the enclosure. Furthermore, the operations are also performed within the enclosure, physically separating the operator from the source of exposure. This is a paradigm shift in control philosophy relative to the previous hoods we have mentioned. For snorkel trunks (or any traditional capture hood, for that matter) and downflow booths, the emission source occurs within the same airspace as the operator. Therefore, strong air currents are required to maintain containment and provide employee protection. With VBEs, the source material and physical manipulations all occur within the enclosure.

Since the material is handled within the enclosure, significantly less air is required to ensure containment. For many LEV capture hoods, such as walk-in booths or slotted hoods, the required airflow to achieve containment can be over 2,500 CFM depending on the factors we discussed in Chapter 6. For VBEs, they have been engineered to achieve desired containment levels with airflows ranging from 500 to 800 CFM, largely depending on the size of the VBE. These reduced airflow requirements have a few

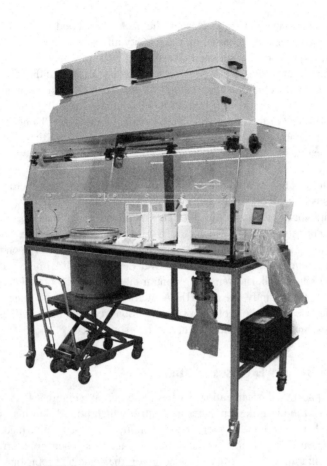

FIGURE 7.4   An example VBE. Courtesy of Flow Sciences.

ancillary advantages. First, less air required to achieve containment means less makeup air is needed, so the organization saves money on utility costs. Second, the lower airflow rates also translate into reduced velocity through the working space, causing fewer disruptions (but becoming more susceptible to cross-drafts). The utility requirements for VBEs are also considerably less than capture hoods as well. Most VBEs are well-designed and have undergone extensive revisions over the years. As a result, they are rather robust systems, often possessing a Ce between 0.8 and 0.85. With such efficient conversion rates, less SP is typically required for VBEs than other hoods. For this reason, they are also popular choices for retrofit projects which may not offer a significant amount of SP at the point of operation.

Despite all the advantages that VBEs offer, they also suffer from notable disadvantages as well. Notably, they are typically restricted to operations involving small quantities of materials, normally a few kilograms at most, but usually on the order of several hundred grams or less. Larger quantities of materials, especially bulk material, tend to overwhelm the laminar air streams and break containment. Likewise, only low-energy processes can be performed within VBEs. Since the airflow is laminar and relatively low velocity (often on the order of 75 ft/min), any processes which imparts significant energy/velocity to materials can break the containment barrier and expose operators. Another disadvantage, which is not unique to VBEs, is that they are subject to operator performance. Because the face of a VBE is open, operators can freely bring contaminated objects in and out of the VBE, the most common of which is gloves. Significant training is required for operators to properly use the enclosure and overcome the tendency to remove soiled gloves from the enclosure. This is especially true for handling potent compounds. Many organizations work around this drawback by requiring operators to wear two layers of gloves and to remove the dirty, outer layer prior to exiting the enclosure.

7.2.4 GLOVEBOXES/ISOLATORS

Gloveboxes are what most operators mentally conjure when they first hear about containment (Figure 7.5). They are the classical means of traditional containment and are still the gold standard for small-scale operations. Typically, these engineering controls are constructed of 304 or 316 stainless steel to provide a smooth, non-porous surface with a rigid skeleton for superior containment performance. Properly designed and constructed gloveboxes can readily provide containment levels in the single nanogram per cubic meter range, while superior devices have been shown to achieve levels even lower, in the sub-nanogram per cubic meter range of containment. These devices are also highly customizable, with almost infinite variations being devised within the industry to accommodate myriad of processes and tasks. However, most of these, if not all, rely on typically small-scale operations with handling no more than a single drum of material at a time (upward of 20 kg or so, but typically far less than this amount). The containment levels offered by these devices make them ideally suited for handling potent active pharmaceutical ingredients, or HPAPIs. Such materials are typically classified as OEB 5 or higher substances. In Chapter 3, we mentioned the antibody-drug conjugate (ADC) industry and their capacity to handle the potent "warheads". Those substances, including various auristatins and duocarmycin analogues, are the prototypical substances that would be handled within gloveboxes.

As with VBEs, gloveboxes have the added benefit that any open handling of material is performed with the exclusion of the operator; that is, drums, bags, and other emission sources are isolated away from the operator. But the added front glove port provides significantly enhanced protection for the operator in comparison to the VBE. Because these units are completely sealed, there is less of a

FIGURE 7.5 An example of a hard-walled glovebox. Courtesy of Flow Sciences.

concern regarding other aspects of LEV design such as Ce and required SP for the system. While these are still important, they don't pose as critical a role during the design phase since they typically require significantly less SP to make them operate. However, all the other downstream aspects to LEV system are absolutely vital: the ductwork must be sized appropriately to ensure that the potent materials do not settle out from the air stream, the air-cleaning device must be incorporated with a safe change (bag-in-bag-out) system to protect employees during routine filter changes, and the fan powering the system must be sized and maintained appropriately so as to ensure continuous containment.

As it would be expected, movement within gloveboxes is very restricted. Not only are the operators bound by the physical limitations that the glove ports provide, but the rigid shell of the box itself limits the amount of working space. Consequently, a frequent complaint among workers stem from ergonomic woes: awkward bending, over-reaching or over-extending, and gloveboxes placed too high or too low are common issues. The small working area also restricts the amount and type of work to be conducted within them. It is not unusual for a glovebox to be dedicated to a single task, such as weighing, and then requiring the material to be passed to another glovebox for further processing. In this manner, multiple gloveboxes are frequently found "coupled" together by means of pass-thru ports or contained transfer protocols (such as with rapid transfer ports, or RTPs). It is for these reasons that traditional gloveboxes are restricted to smaller scale operations.

The success of gloveboxes showcased the power of full containment. For potent compounds, the danger is not restricted to handling the raw materials at the beginning of a workstream. Rather, all downstream operations which utilize HPAPIs have employees who can be exposed to such materials. The concept of isolating these processes using the same principles of gloveboxes has opened an entire industry for the biopharmaceutical field: process isolators. In this fashion, glovebox-type containment options are designed for all aspects of pharmaceutical operations, including milling, mixing, transfers, tablet presses, packaging, filtering, drying, centrifuging, and more. For HPAPIs, it is not uncommon to have entire rooms be wall-to-wall isolators which contain all equipment for the workflow. As it would be assumed, such buildouts require extensive planning and can incur significant capital investment.

7.2.5　Flexible Containment

Within the last several decades, a new take on containment has gained a massive following in the pharmaceutical realm: flexible containment (Figure 7.6). Flexible containment is an increasingly popular means to achieve containment. All of the previously discussed means of containment have drawbacks that are unique to those design types, but they also share another significant drawback – cleaning requirements. In the pharmaceutical industry, operations almost always occur under current good manufacturing practices (cGMP) requirements. One of the critical components of cGMP, or current good manufacturing practices, is cleaning. It is imperative that work areas and equipment be cleanable so as to prevent cross-contamination between batches and different products (SKUs). These requirements apply to containment devices and any equipment they may contain as well.

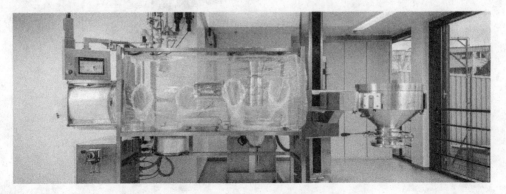

FIGURE 7.6　An example of flexible containment in the form of a flexible ADC isolator. Image courtesy of Lugaia AG Containment Solutions.

Consequently, all the hoods we have discussed so far must undergo extensive cleaning, often through a validated process to ensure repeatability. These cleanings typically stem from the quality requirements, but they also serve an important health and safety consideration to prevent employees from surface exposures. All in all, cleaning requirements are often tedious and lead to significant downtime. Flexible containment largely bypasses this requirement of cGMP cleaning. The flexible containment devices are intended for one-time use and are typically disposed of as hazardous waste upon completion of a process, bypassing the need for cleaning validation altogether. There are considerable advantages to this route. The first is cost. As we have discussed for the previous containment devices, there can be a significant capital investment that is required. These costs only increase as additional customization is added, including all stainless-steel fabrication, equipment integration, filtration requirements, etc. Flexible containment devices have a much lower capital threshold which is often seen as appealing to line managers. Another significant advantage is the time savings. Since cleaning and change overs between batches or products is often time consuming, being able to simply dispose of the work environment vastly expedites the change over process. There is also significant time savings regarding the cleaning validation process itself, which can often be laborious.

Flexible containment technology has also increased its influence within the sector; that is, they are not restricted to solely being a flexible version of a glovebox. It has variants in forms ranging from continuous liners (used for disposal of waste materials in other containment devices and also in the packaging of hazardous substances into drums and flexible intermediate bag containers) and single-use discharge bags for potent compounds. This increase in flexible containment utility is perhaps linked to the displayed efficiency of containment itself. Modern examples of this technology, particularly for isolators, have demonstrated consistent containment abilities in the sub-100 ng/m^3 range, making them prime candidates for handling of HPAPIs. Some flexible containment devices have even demonstrated containment levels lower than 10 ng/m^3, although this may be dependent on the amount of material handled, the speed of the process, and the particle size of the material being handled. Furthermore, these levels of containment have also been realized with the other variations of the technology, making it even more robust in the industry. The ability to adequately protect personnel while removing time-consuming manufacturing steps makes flexible containment a highly attractive option for many managers and industrial hygienists alike.

The drawbacks to flexible containment should also be noted. The first is cost. While we previously mentioned that flexible containment has a lower initial capital cost than rigid isolators, the systems themselves are not without their price tag. For organizations that run a wide range of batches and products on a fast timetable, the amount of flexible containment that could potentially be required can add up quickly. In such circumstances, it is probably more cost effective to invest in rigid systems and undergo the cleaning validation steps as it would save money in the long run. Additionally, if large quantities of flexible containment are utilized, it essentially creates a whole new unique waste stream that must now be disposed. Since these units are composed of single-use plastic, they are most likely incinerated. In today's environmentally conscious society, most pharmaceutical organizations are trying to minimize their use of single-use plastic. Even though single-use plastic is vital to the industry and cannot be eliminated outright, processes which considerably (and potentially unnecessarily) add to this volume will almost surely be avoided. Flexible containment sees significant benefits for processes which have infrequent batch runs and with contract development and manufacturing organizations (CDMO) which often have completely new and varied projects.

Despite these drawbacks, the demand for flexible containment continues to grow. More and more companies are entering the market to offer competitive products as well as creative takes on old variants.

7.3 CONTAINMENT OPTIONS FOR HANDLING LARGE QUANTITIES OF MATERIALS

For large-scale batch runs, significant quantities of starting materials are needed. Bulk materials can be provided in various packages including bags, drums, and FIBCs. The following sections are an abbreviated listing of commonly utilized containment strategies for handling bulk materials in the pharmaceutical industry. Like the devices highlighted in the previous section, this list is non-exhaustive as there are numerous types of devices and almost infinite variations of those that are listed here. However, it is hoped that the items presented below provide the reader with a good foundation upon which to build for further research and reading.

7.3.1   Bag Discharge Station

One of the most common forms of bulk material availability is in the form of bags. They are easily filled, shipped, stored, and handled. However, they do not always cooperate in terms of material discharging. A common occurrence with bags is the significant amount of material that does not make it into a reactor vessel when they are opened and poured. When handled in this manner, personnel can be exposed to high levels of materials. Containment becomes almost a necessity if these operations occur frequently.

The bag discharge station is a common containment solution for bag emptying operations. Consisting of a semi-enclosed laminar flow hood placed around a discharge spout directly above a tank or hopper, discharge stations offer a convenient option for dispensing bags of material. The hood is frequently directly connected to an LEV system for dust collection but also contains standalone HEPA filters to collect dust. The bags often rest upon a grate, and when opened, the material falls through the grate and into the tank. There are numerous variations of this design, including offering remote discharge options in which the discharge station is located some distance away from the processing point. In these situations, the bulk material is accumulated in a hopper and then conveyed through piping into the ultimate destination. Such arrangements are common where space is a significant constraint. However, the overall distance from the discharge station to the tank is another constraint that must also be considered as well. Other variants of the design include the incorporation of material grinders or delumpers. These devices break up large pieces of the starting material into smaller, more manageable pieces. This is especially prudent for bagged materials which tend to harden upon stagnation or upon exposure to moisture. Other situations which delumpers are useful include altering the rate of dissolution into a solvent or enhancing the flowability of a material through conveyance piping.

Because bag discharge stations are semi-enclosed (that is, the front face is typically open while baffles are present on the sides), there is often concern about the exposure potential to operators. Many designs are available which have undergone surrogate powder testing, and well-designed discharge stations can achieve task-based containment as low as 100 µg/m³. Of course, such performance is largely dictated by operator usage, particle size, and rate of discharge, but surprisingly effective containment can be achieved with these devices. The containment performance of discharge stations typically relegates them to materials belonging to OEB 1 or 2, or substances that are typically considered less hazardous than more potent materials. Such restrictions are often of no consequence, since a lot of bulk materials used with bag discharge stations are excipients.

A worthwhile note regarding modern bag discharge stations is the frequent addition of a ventilated bag compactor. An often overlooked source of exposure for operators occurs during the opening of raw material containers and the closing and/or disposing of "empty" containers. When bags or drums have undergone discharging, the parent container is not 100% empty; rather, there is often still copious amounts of residual material to which operators can be exposed. Often the closing of drums or the act of discarding emptied bags can lead to unintended exposures. Manufacturers of bag discharge stations have taken these situations into account by installing ventilated bag compactors. Once a bag is discharged into the hopper, operators keep the empty bags inside the containment device and slide the bag laterally into the compactor. In this manner, the bags are disposed and the operators remain protected due to additional containment.

A drawback for bag discharge stations is the volume of air that is often required. To be effective, many stations require a volumetric airflow in the neighborhood of 2,000 CFM to achieve laminar air flows of 100–200 ft/min at the operator interface. The exact volume is dictated by the size of the station and the desired capture velocity of the hood. But as we mentioned in the last chapter, all the exhausted air must be replaced with conditioned makeup air, adding to the overall cost of the device. A second drawback is difficulties when multiple products must be manufactured. For cGMP purposes, different products require that components which come into contact with materials be completely clean and free of contamination. Achieving this with bag dump stations is very difficult and becomes even more difficult if conveyance lines are involved. To circumvent this, bag dump stations are typically dedicated to the discharging of a single material which is then stored in a day tank. If different products require varying amounts of the material, the production line can utilize different amounts from the day tank rather that directly from the bag discharge station. However, this situation is not always feasible if a large number of raw materials are needed for variety of products. A final drawback is often the ergonomic factor. For bagged materials, the most common size encountered commercially is 25 kg (although different bag sizes do exist). A process which requires a substantial amount of material may utilize dozens of bags. Operators physically handling and manipulating 25 kg bags can encounter MSDs fairly quickly. Workarounds to this dilemma involve the use of bag lift devices.

Regardless of these drawbacks, bag dump stations are highly popular and effective containment devices. Well-designed stations are fairly efficient, with Ce values in the neighborhood of 0.8. Consequently, the amount of SP required at the hood is often in the range of 2.00 in. w.c. Depending on size constraints and material versatility, bag dump stations are often candidates for line upgrades or retrofit projects.

7.3.2 Ventilated Collars

Countless operations involve pouring, discharging, or otherwise utilizing gravity to move material from one piece of equipment to another. In the past if these operations were not done via enclosed piping, the operation was often done in an open environment. Depending on the height of the activity, significant energy can be imparted to the material, thereby causing large plumes of material. Consequently, these activities led to unacceptably high levels of exposure. Yet for many of these operations, totally enclosing the material stream cannot be reasonably accomplished for various reasons (cleaning validation between batches primarily among them), and traditional capture hoods are typically ineffective at capturing the amount of dust needed to effectively manage the exposure risk.

One solution that has gained popularity in recent years is the use of a ventilated collar (Figure 7.7). Essentially, ventilated collars are round LEV capture hoods through which discharge operations can be performed. When supplied with sufficient airflow, the collars provide ample capture velocity for many tasks which would otherwise be left with few options.

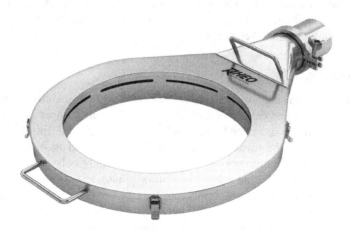

FIGURE 7.7 A ventilated collar. Courtesy of Rheo Engineering.

The effectiveness of the collars is largely dictated by the overall design and implementation of the discharge apparatus relative to the collar. If a mixing tote containing material were to be discharged into a hopper through a ventilated collar, but the discharge point is positioned well above the capture zone, then the collar will be ineffective. For most operations involving large amounts of material, the discharge point (where the bulk material and dust leave the parent container) must be below the capture zone. In this manner, any plumes that are generated in an upward trajectory travel through the capture zone and therefore containment is achieved. Ventilated collars have been successfully deployed in filling operations with hoppers, centrifuges, mixing vessels, and even drums. Other advantages of ventilated include superior cleanability between batch runs. The collars can be disassembled with ease and thoroughly cleaned, thereby appeasing even the most stringent quality groups. This singular mark makes them appealing to organizations dealing with a large number of bulk materials since ensuring quality between batches is paramount.

As would be expected, the containment level offered by ventilated collars is highly variable. In carefully controlled testing facilities, containment levels below 100 μg/m^3 can be achieved, sometimes as low as 50 μg/m^3. However, in real-world applications, the actual containment level can vary considerably. As usual, this is entirely dictated by the usual factors: particle size, amount of material, rate of transfer, and the distance from the emission source (dust plume) to the capture zone. In manufacturing scenarios, time is often of the essence, so processes are typically performed much more quickly than in test scenarios. In these settings, ventilated collars can routinely provide containment between 100 and 500 μg/m^3. While these levels may not immediately sound appealing, the duration of exposures for such operations are typically short-lived, meaning the overall TWA for exposures can still be achieved with ease. Ventilated collars are often relegated to OEB 1 or 2 materials due to these containment levels. However, this can be further enhanced with respiratory protection or additional engineering controls. Organizations often supplement the use of ventilated collars with a requirement for respiratory protection as well, usually N-95 respirators. Additionally, operations involving ventilated can be combined with other previously mentioned control solutions such as downflow booths.

As with bag discharge stations and other hoods, the primary disadvantage of ventilated collars is the amount of air required to work efficiently. The required volumetric airflow depends on the size of the connection to the exhaust system as well as the diameter of the collar itself. A common setup is to use a 16-inch diameter collar with a 4-inch exhaust connection. In this arrangement, upward of 500 CFM is required to achieve the required capture velocities. While this may not seem like a significant volume of air compared to other hoods discussed thus far, the SP requirements to achieve this flow rate are considerable. Ventilated collars often have a Ce between 0.6 and 0.7. Thus, for the ventilated collar setup described a requirement of 5.00 in. w.c. of SP would need to be fulfilled. If ventilated collars are being installed as part of a retrofit project, it is imperative that the engineers know how much SP is available at the connection point. Otherwise, the desired containment will never be achieved. Despite this disadvantage, ventilated collars are growing in popularity due to their ease of use, installation, and versatility.

7.3.3 DRUM DISPENSERS

Safely and efficiently dispensing bulk materials from drums is a longstanding problem in virtually every industry, the pharmaceutical sector included. A wide variety of operations utilize materials from drums, and these materials can be liquids (such as solvents) or solids (powders, granules, crystals, etc.). Liquids often have the advantage of being able to be pumped out of drums through sealed piping, but solid materials often do not enjoy such benefits. Many bulk drum dispensing options stem from the same concept: Pick up a drum and allow gravity to perform the work of emptying the drum. While this option works well in principle, the practice of having solid material from any height present an unnecessary exposure risk for operators. Plumes of dust are common occurrences for these scenarios. Fortunately, advances in containment have been applied to drum dispensing operations.

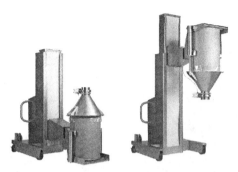

FIGURE 7.8 An example of a drum inverter for the dispensing of solid materials. Courtesy of Rheo Engineering.

Figure 7.8 shows modern applications of a drum inverting process. In the image, a funnel is placed over an open drum, creating a seal which prevents bulk powder and vagrant dust from leaking out. The drum is then fully inverted 180°. The funnel is outfitted with a valve which controls the flow rate of the material from the drum. Additionally, the spout of the funnel can be connected to the desired transfer point through hard piping or flexible sheeting. In this configuration, complete dust containment can be achieved, with levels often seen around 1 μg/m^3 or less in some instances. An alternative is to have the material freely dispense from the funnel into the desired receptacle in an open environment. As one would expect, this significantly increases the likelihood of dust exposure and containment levels are decreased. Frequently, if open dispensing is performed, an additional engineering control (downflow booth, ventilated collar) is utilized or the requirement for respiratory protection is instituted. This drum inverting method has advantages of taking up less room within a facility but more importantly, the application can be applied to a wide variety of materials. The funnel can be easily cleaned and used for a multitude of products.

Other drum dispensing methods employ a similar concept of utilizing gravity to dispense the material. However, a key difference is that the point of dispensing is removed from the general work area within a ventilated enclosure. The dispensed material can be transferred to a hopper or another vessel, or it can be conveyed pneumatically through piping to other locations where it is needed for manufacturing. By enclosing the dispensing point, dust levels are drastically reduced if not outright eliminated altogether. The enclosure achieves dust removal by being directly connected to an LEV system. The disadvantages of such systems are that since the system is fully enclosed, cleaning is difficult. Consequently, such systems are often established for a dedicated manufacturing line that produces a single product, or it can be used for a material that is common to a multitude of products. In this manner, the material is dispensed from the drums and transported to large holding tanks such as silos where it remains until a required amount is needed and called for by the manufacturing system. Another disadvantage is the potentially large volumes of air that are required to achieve the level of dust reduction.

The containment levels offered by large bulk drum dispensing operations such as those highlighted in this section are suitable for handling materials that are typically less hazardous, such as OEB 1 or 2 materials. Again, this often restricts the materials being handled to excipients, additives, flavors, colors, etc. For materials that are more potent, including APIs, open drum dispensing is simply not an option. Modern approaches for dispensing APIs from drums (OEB 3 and higher materials) have sought to combine drum tipping/dispensing operations with the containment capabilities of gloveboxes and isolators. Figure 7.9 shows an example of this combined engineering control, often referred to as a material transfer station (MTS). The drum containing the API is placed on a drum lift which brings it into the enclosure via a port in the back. The port can be sealed with an inflatable bladder to create a dust-tight seal. Once brought in, the drum can be opened and the material manipulated by an operator by using the glove ports. Depending on the equipment setup

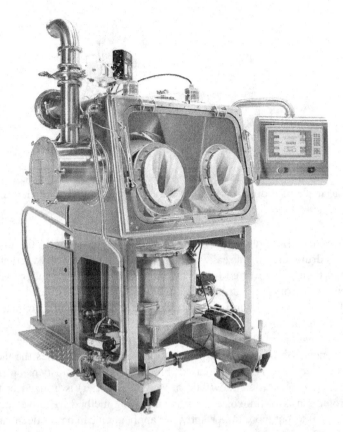

FIGURE 7.9 An enclosed drum dispensing material transfer station (MTS) suitable for potent compounds. Courtesy of Rheo Engineering.

and specifications, the bottom of the isolator can discharge directly into a tank, vessel, or a hopper which can then convey the material to the next phase of the process. Several additional features are unique to the equipment, including the use of HEPA filters for exhausted air before it enters the LEV system and the implementation of continuous liners to dispose of any contaminated drum liners. The units are also customizable, with the ability to integrate scales, delumpers, and clean-in-place (CIP) technology to satisfy both quality and safety personnel.

When used as designed, the containment capabilities of this equipment often rival those of traditional gloveboxes. Containment performance of such devices exhibit routinely in the range of 0.05–0.2 µg/m^3. The performance of these units is dependent upon how well the operators use them, the rate at which drums are emptied, and how well the units are maintained. There are many potential leak points around an MTS, such as the glove ports, waste chute, and even the drum port. Ensuring adequate preventative maintenance is exercised on these devices goes a long way to maintaining the level of containment required for handling potent compounds on a larger scale.

7.3.4 FLEXIBLE INTERMEDIATE BULK CONTAINER (FIBC) DISPENSING

While bags and drums represent a common means of providing bulk material for a process, some operations require still larger quantities of material. In an industrial manufacturing setting, the largest scale of material availability is typically provided via the flexible intermediate bulk container, or FIBC. Sometimes referred to as "bulk bags", the FIBC offers a cost effective and convenient means of providing raw materials on a very large scale. A common FIBC can provide approximately 2,000 pounds of material in a single bag. A repeating theme throughout this book has been the generally accepted notion that the more material is handled by an operator or used in a process, the greater the probability of excess exposures. Since FIBCs represent the single largest source of materials, it stands to reason that they also offer the greatest potential source of overexposure.

Discharging of materials from FIBCs is a commonly encountered processes in a manufacturing setting. Like dispensing materials from a drum, FIBCs often take advantage of gravity and discharge their materials downward into a receptacle. The receiving device can be a tote, sifter, silo, hopper, or some other container. The significant point of concern is the connection between the FIBC and the discharge spout. If the connection between these two points is not performed properly, then a large amount of powder and dust is likely to enter the general working area. Contemporary FIBC discharge stations take advantage of the fact that most bulk bags have two distinct plastic liners within them, an inner and outer liner. Prior to opening the FIBC, the operator places the inner liner within the discharge spout, which leads directly to the receiving receptacle. The other liner, on the other hand, is placed around the outside of the discharge spout. At this point, an inflatable bladder expands outward from around the exterior of the discharge spout, forming a dust-tight seal against the outer liner of the FIBC. Once this setup has been completed, the FIBC is opened and the material is dispensed. This particular method of FIBC connection and discharge has been highly effective at mitigating large dust concentrations in manufacturing settings.

Yet even with this advancement in FIBC connection, there can still be fairly considerable powder build up around the spout. During disconnection, operators can be exposed to the powder which did not make it down the discharge funnel. To assist with this, some FIBC discharge stations have incorporated additional containment around the connections. Small compartments with a sealable door are placed around the FIBC discharge connection provides an additional barrier between the operator and the operation. Importantly, these compartments have dust collection capabilities by being connected directly to an LEV system. The incorporation of dust collection to the FIBC discharge system provides still yet another layer of protection for the operators. The airflow requirements for the localized dust collection can be as high as 1,000 CFM, but the airflow only operates during the dispensing task, which greatly minimizes the amount of makeup air. Furthermore, since the area being evacuated is a very small and confined area, the total volume of air being removed is substantially lower than for other LEV systems. Consequently, the required makeup air is also much lower. An additional feature on some of the more advanced dispensing systems include a bulk bag collapser. At the end of a dispensing operation, "empty" bulk bags are not completely empty. The bag collapser feature applies a vacuum to the inside of the bag, removing a significant portion of remaining material.

These layers of engineering controls for operator protection (inflatable bladder, sealable compartment, localized dust collection) have made for significant improvements in the work areas for operators where FIBCs are handled. Typical containment levels experienced in these settings vary widely, often from as high as 1,000 to 500 $\mu g/m^3$. This is largely due to the fact that in manufacturing settings, large numbers of FIBCs are frequently handled. Conceivably, a manufacturer could go through as many as 20 bulk bags each day. That means upward of 40,000 pounds of material is being handled/dispensed each day. With that much material being processed, the exhibited containment levels are actually quite impressive. Nevertheless, the containment capabilities for such setups restrict the process to largely innocuous materials. Typically, substances such as excipients and fillers are dispensed from FIBCs.

Yet there are some instances wherein more hazardous or potent materials need to be dispensed from an FIBC. Just as bag discharge stations and drum dispensing stations have incorporated glove box technology, so too have bulk bag discharge stations. Some operations can require an API or a more hazardous material to be delivered and discharged via FIBC. In instances such as these, bulk bag connections are made within a glovebox variation of the previously described discharged apparatus. The same safety mechanisms and layers of engineering protection that are present in open FIBC discharge operations are also present in the glovebox variant but with the added component of isolation from the operator. Excellent containment can be achieved with these setups when used appropriately.

7.4   ASSESSING CONTAINMENT PERFORMANCE

The abbreviated survey of hoods and containment devices shown in this chapter all speak to various levels of containment performance. Each hood has a general range to which containment can be achieved, but this performance is largely dictated by the amount of material being processed, its physical properties including particle size and dustiness, the frequency of the operation, and the amount of airflow provided to the unit. Each type of containment device is designed to perform to a certain level with a specific process, and deviations from these design specifications can drastically affect the containment level offered by the unit. But how exactly is the containment performance of a hood verified? This section will briefly cover some of the strategies that industrial hygienists use to assess verify containment performance.

7.4.1   Assessment of Fume Hoods

When most people think of hoods, they typically envision laboratory fume hoods, covered earlier in this chapter. Fume hoods are where the bulk of small-scale operations take place, including initial research and development activities. Yet fume hoods are not just restricted to medicinal chemistry-type activities. Indeed, process chemistry scale-up activities can also occur within fume hoods. These may also include not just typical fume hoods but also floor-mounted hoods. The primary goal of these controls is to minimize operator exposure to potentially hazardous vapors and gases from solvents and other sources.

To ensure these control devices are operating as intended, fume hoods often undergo evaluation using criteria from the ASHRAE 110–2016 standard, "Methods of Testing Performance of Laboratory Fume Hoods".[2] This document has been adopted as the industry standard for assessing containment performance in fume hoods. It utilizes both qualitative and quantitative means of assessing a hood. From the qualitative viewpoint, the standard requires the use of smoke generators to view the airflow into the hood but also how the air moves within the hood itself. The quantitative aspect is more involved. The standard requires the assessor to measure face velocities using an anemometer within an imaginary grid along the opening of the hood. Multiple measurements are required, and an overall average is calculated. Next, a tracer gas, stipulated within the standard as sulfur hexafluoride, is emitted in a controlled release at a set distance within the hood. A mannequin equipped with a detector within the breathing zone is placed in front of the hood at a typical operating position and distance. The detector measures the amount of tracer gas that escapes from the hood.

The techniques specified within the standard have been applied to countless hoods throughout the industry, and the test data are often available from manufacturers for "off-the-shelf" hoods. Moreover, the process can be used for fume hoods that are already in use within a facility. But several questions immediately arise and are often asked during the process: What constitutes a passing test? And how high should the fume hood sash be placed for the test (for vertical sashes)? For the first question, there is no straight forward answer. Results are often reported as the average ppm or ppb level of SF6 that was detected in the worker breathing zone, but they can also be presented as a percentage of the amount that was present within the hood. For the test designer, the passing criteria must be decided upon prior to initiating the test. For instance, the site EHS representatives (or the organization) may have a requirement that no more than 150 ppb of tracer gas be detected outside of the hood. Alternatively, the requirement may be that no more than 0.05% of the concentration of tracer gas within the hood be detected outside of the hood. While these values presented are arbitrary, there is no standardized "passing" score for an ASHRAE 110 test. The desired performance of a fume hood is one that must be decided upon by the risk team and the organization in general. The level of performance may be dictated by the materials to be handled within the fume hood, that is, based on a control banding scheme; however, this is extremely difficult to implement since activities within fume hoods vary by project. And since R&D and medicinal chemistry projects experience considerable turnover and uncertainty, predicting exactly what a fume hood will house and what the anticipated OEL will be.

Regarding the second question on sash height, this is perhaps the most often questioned aspect of a fume hood test if it is performed on an existing hood in the lab. Moreover, there are a few terms that need to be identified. "Design height" is the height of the sash that the manufacturer of the hood anticipated while designing the hood. Depending on manufacturer, the design height may be the factor for determining the total airflow into the hood. "Operating height" is the placement of the sash that an operator will typically place it for comfort purposes during operation. This may not coincide with the design height and is often dictated by ergonomic factors such as operator height. "Maximum height" is when the sash is lifted as high as it can go. Such placements occur frequently, especially during equipment set up and tear down, and many fume hood users automatically fully open a sash when working in a hood. These differences may seem subtle or even inconsequential, but it gets back to the notion of engineering control performance being based on how the control was designed and intended to be used. A manufacturer may design a hood based on a given flow rate and an 18-inch sash height, but realistically operators may not abide by this unless there are additional preventative controls. To account for this, the ASHRAE 110 standard recommends performing the tests with the sash height at the maximum level. While this ensures a passing test during a "worst case scenario", overall, it means that a much larger volume of air must be exhausted. This is due to a larger cross sectional surface area through which air passes (recall that volumetric airflow, Q, is equal to the air velocity, V, multiplied by the area, A). Consequently, more makeup air will be required to balance out the building, and additional energy costs will be incurred. For locations that utilize a large number of hoods, such as a large R&D center, these costs add up quickly.

Another point to raise regarding fume hood testing is that the majority of these are performed as part of a factory acceptance test (FAT). These are done in controlled environments and, importantly, with empty hoods. The fact that the hoods are empty is significant because once installed, users tend to put equipment within them. In a laboratory setting, the types of equipment found in a fume hood can range from a simple stir plate to large, bulk items such as HPLCs, pumps, rotary evaporators, and ovens to name a few. All of these items disrupt the airflow patterns within a hood and can significantly impact the performance of a hood. It is entirely possible that operators will complain of odors from hoods containing equipment, or industrial hygiene sampling will give consistently higher results for personnel using such fume hoods. If an ASHRAE 110 test is performed, the hood may not meet the necessary criteria to pass the test even though it passed during the FAT and has the correct volumetric airflow and SP being delivered to it. This is especially true for floor mount hoods where process chemistry equipment is utilized.

This is an instance wherein there is a stark difference in how the containment is assessed initially versus a real-world use. It is possible that the ASHRAE 110 test for an "in use" hood may need to have the results calibrated, so to speak, to account for any differences. It is important to take the whole performance of the unit into account. Along with the tracer gas testing results, the IH should also evaluate actual industrial hygiene sampling results from actual processes. If the tracer gas test did not perform as well as it did during the FAT but the sampling results still indicate that operators are likely to be an AIHA Exposure Category 0 or 1, then the hood is still successful and the risk is successfully reduced for the users. In situations such as these, the results from the "in use" tracer gas test should be used as the baseline target for future hood assessments, and not necessarily the results from the ideal laboratory testing protocol.

7.4.2 ASSESSMENT OF CONTROLS FOR POWDER HANDLING

While fume hoods have benefited from the existence of a uniform standard against which their performance can be assessed, containment devices specifically for handling powders and dusts did not have such a standard for a long time. Yet the requirement for powder containment has increased exponentially within the pharmaceutical industry, in particular for equipment in which HPAPIs are to be used. Reliable methods to assess particulate containment efficiency were needed to verify performance of hoods before they were installed in the workplace.

To meet this need, the International Society for Pharmaceutical Engineering (ISPE) published the *Good Practice Guide: Assessing the Particulate Containment Performance of Pharmaceutical Equipment*.[3] While not an internationally recognized standard in the strictest sense, the guide was authored by a collective of containment engineering experts, industrial hygienists, toxicologists, and other pharmaceutical stakeholders. The intent of the document is to provide standardized guidance on measuring how well a piece of equipment performs under the challenge of powder handling. The guidance is based on years of field experience and sound, recognized practices in industrial hygiene.

Containment of powders is assessed by performing the anticipated task in the appropriate manner while utilizing the desired engineering control. During the test, calibrated sampling pumps are set up at likely leak points around the control device to measure for any powder or dust emissions. The placement and height of each sampler is dictated by the needs and design of the equipment being tested, but the ISPE provides guidance as to best practices for repeatability. In addition, the operator performing the test can also be outfitted with a personal sampling pump in his or her breathing zone. Since each piece of equipment being tested is inherently used for a different task and utilizes varying amounts of material for said task, each powder containment test is unique and individualized. This is in stark contrast to ASHRAE 110–2016 in which the test is essentially performed the same way each time with each hood so as to be able to directly compare results. For the powder containment testing, accurately crafting a sampling plan to acquire the most accurate and useful data is critical. Once again, understanding the process and how it is performed as well as knowing the equipment is vital for the hygienist.

Powder containment testing is often performed at the facility in which the containment device was constructed within a clean and controlled test environment. Often referred to as an FAT, the testing is performed during a mock use of the system. Multiple runs of the operation/task are performed with samples being acquired each run at each anticipated test point. The mock process should include as many potential real-world variables as possible, such as equipment and personnel. Again, this is in stark contrast to ASHRAE 110. The number of runs for the testing should be decided upon early but typically at least three runs are performed so statistical evaluation of the results can be performed.

Since testing should not be performed with hazardous substances that could cause harmful health effects if containment failed, the tests are performed with the use of surrogate materials. Surrogate materials have the many commonalities among them, such as being non-hazardous, availability in bulk, and relatively inexpensive to purchase. The more commonly used surrogates for containment testing include lactose, mannitol, and naproxen sodium, but other acceptable surrogates can include riboflavin, sucrose, and acetaminophen.

Two other vitally important features of surrogates are the reporting limits and the associated particle size. All surrogates enjoy extremely low reporting limits; that is, even very minute traces can be quantitated if they escape containment. Lactose, perhaps the most commonly used surrogate, has a reporting limit of approximately 2.5 ng. Naproxen sodium has an even more sensitive reporting limit, as low as 0.05 ng. The specific reporting limit is dictated by the laboratory performing the analysis and the practicing industrial hygienist should always consult with the lab prior to executing the analysis. But the need for the low reporting limits has never been more important.

For many containment devices, the use of HPAPIs and their extremely low OELs makes the need for extreme sensitivity an absolute necessity. Consequently, for many glovebox applications, the only acceptable surrogate to use is naproxen sodium. This feature becomes even more apparent when the concept of task duration is taken into account. Surrogate powder testing evaluates emissions at a task-based level. Since OELs are typically based on an 8-hour exposure period and surrogate testing evaluates task-based emissions, those performing the assessment cannot directly compare surrogate test data to the OEL. Furthermore, containment emissions do not directly correlate to operator exposure. Yet a point of comparison is still needed to know whether or not the device "passes" the testing. The ISPE recommends the use of a "containment performance target", or CPT, to evaluate a particular engineering control. The CPT is often based on the OEL of the material being handled in the device. If multiple agents are to be used in the control, then the material with the most stringent OEL is used as the basis for the CPT. There is no set protocol for crafting a CPT. It is entirely dependent upon the organization to establish a protocol for crafting CPTs. Some organizations simply use the OEL or a fraction of the OEL, such as 1/10th of the strictest material being handled, but the manner in which the CPT is devised must be decided upon by the organization. Since the downstream ramification is for a device designed to protect personnel and therefore mitigate risk, the CPT values are often risk-based and therefore conservative. Regardless of how they are crafted, the CPT represents a task-based static emission threshold for the equipment.

In contrast, data acquired from operator breathing zones cannot be compared to the CPT. Operators move about during operations, and consequently, their overall exposure will vary considerably. But since testing is task-based, there are no levels for direct comparison. To circumvent this, organizations often craft "designed exposure levels", or DELs, specifically for surrogate powder testing. These are airborne concentrations based on the OEL that take presumed task duration into account. The DEL can be a fraction of the OEL or can be set at the OEL itself (setting the DEL at the OEL can be advantageous since tasks do not take an entire 8-hour shift, therefor allowing periods of zero exposure in between; therefore, if the measured levels are below the DEL, then the likely 8-hour time weighted average will be below the OEL). Just as there is no set method for establishing a CPT, there is no set method for setting a DEL. The manner in which a DEL is set is entirely at the discretion of the organization but must be decided and agreed upon by all parties prior to executing the test. Regardless of how these values are set, their low values necessitate the need for a surrogate with very low reporting capabilities.

While analytical sensitivity is the most significant variable when choosing a surrogate, it is not the only variable to consider. Particle size distribution is another critically important factor as well. We have mentioned the importance of particle size in previous chapters, and its importance is showcased yet again. Materials handled in operations with larger particle sizes often have a harder time escaping containment (when handled properly) than smaller particle sizes. It is important to properly simulate not only the task itself but the materials being handled. If a process involves materials that are granulated or pellets with average particle sizes of 250 µm or larger, it would not be prudent to use a surrogate which is micronized with an average particle size of 10 µm or less. Attempting to match particle sizes gives an approximation of the "dustiness" of the operation and therefore gives more accurate and realistic results to the real-world application.

Once a surrogate has been selected, a sampling plan crafted, and sampling performed the next critical piece of information is knowing whether or not the device passed the testing. The passing criteria for a surrogate powder test is, again, dictated by the organization. Many pharmaceutical companies utilize the rules and recommendations found in the European standard EN-689,[4] but this is not an absolute requirement. Since the data are industrial hygiene data points, traditional statistical tools can also be utilized to determine if the 95th percentile of the collective data exceed the CPT for static emissions (and the DEL for the breathing zone samples of operators). Yet a common occurrence for devices is the acquisition of censored data points, and traditional statistics are difficult to perform with censored data. In contrast, the use of Bayesian statistics can circumvent this shortcoming while still providing a measure of where the 95th percentile lies in relation to the CPT. This approach is being adopted more frequently within the industry as it is straightforward and is consistent with the use of BDA for routine industrial hygiene analysis.

It is important to keep in mind that surrogate powder testing also requires the hygienist to record airflows into the testing device (for non-glovebox devices) as well as the SP serving the units. The room conditions must also be recorded and reported, including temperature, relative humidity, room air changes, relative position of personnel and equipment, and any notable cross-drafts which may interfere with testing. Good documentation practices pay dividends in the final report, especially if stakeholders have questions later on.

Surrogate powder testing as described is often performed at test facilities for new equipment (the FAT). However, the same protocol is often performed upon installation at the facility where it will be used. This process, called the site acceptance test (SAT), is often a requirement of the change control process and verifies that the equipment performs the same in the destination facility as in the test facility. The same exact test protocol that was used for the FAT should be executed for the SAT to make sure the results are directly comparable. Additionally, surrogate powder testing can be performed on any piece of equipment that is currently installed at a facility. It is not relegated to only new equipment. The utility and flexibility of the ISPE guidelines allow for a wide variety of equipment to be assessed for their containment capabilities. This is especially recommended for control devices that utilize powders for which validated industrial hygiene sampling methods do not exist. In previous chapters, we highlighted that for substances lacking analytical methods, the usual alternative is to use gravimetric analysis, which is non-specific and has a poor reporting limit. For these instances, surrogate powder testing is a welcomed solution as it can provide significant input to a wide array of processes and control devices.

There are a few notable drawbacks to surrogate powder testing. Primary among these is the cost. While surrogates themselves are relatively widely available, their cost on large scale can add up quickly. For instance, if a large isolator is being challenged and requires the need of naproxen sodium on a multi-kilogram scale, the cost for the surrogate powder alone can be north of $10,000. In addition, for each run there can be a multitude of samples taken, and if multiple runs are performed, the total number of samples acquired can become costly. For example, if the isolator example had seven potential leak points requiring sampling, plus the operator, a total of eight air samples would be collected for each run. If a minimum of 3 runs are performed, then 24 air samples will be submitted to the lab for analysis. While the cost of analysis of surrogate samples is not overtly expensive, the large number of samples can quickly add up. There is also the time factor associated with surrogate powder sampling. To be performed correctly and avoid cross-contamination, the testing takes time, often over the course of multiple days. If the testing is performed for existing devices already in use, manufacturing must be delayed so the testing can be performed. The lost production time should be weighed accordingly when scheduling testing.

Another drawback relates specifically to testing existing containment devices. Most facilities have a large number of such equipment, and to effectively test all of them could easily exceed the sampling budget of even the most heavily financed department. Strategically choosing where to focus attention first is a significant challenge and responsibility of the site industrial hygienist. His or her choice should be appropriately justified based on the potential number of folks who can be exposed and/or the materials being used. Additionally, testing existing devices presents another issue that may pique the interest of the residing quality department. Since most facilities are within a cGMP setting, the introduction of foreign material into a system can have significant downstream effects. The quality group may vocally object to the use of a surrogate within the system without any accompanying cleaning protocol. Such objections are prudent and valid: the hygienist must always remember that the facilities are used to produce medicines that are delivered to patients and altering anything to cause potential deviations and supply chain disruptions must be avoided at all costs. It is important to work closely with the quality department to ensure they give their blessing prior to surrogate testing any equipment within a cGMP area.

7.5 SUMMARY

The risk treatment phase of the risk management process in the pharmaceutical industry really begins with protecting the operators when they are directly handling materials. The engineering controls utilized have the greatest impact in this reduction and should be matched accordingly to the task being performed. Other constraints also exist, such as space and the required level of containment for the materials. An important concept is the idea of containment. Historically, containment almost always mean full enclosures, but this is no longer the case. Containment now refers to the ability of a device to minimize unintended exposures of material to operators and the environment. This chapter presented a very high level and abbreviated survey of frequently utilized hoods or containment devices and how they are used along with containment capabilities. For each device type, there are countless variations, and it would be short of impossible to document all of them. Containment can be objectively measured for engineering controls using standardized methods (for gases and vapors) and industry best practices (for powders). The entire spectrum of selecting a particular engineering control and assessing its containment abilities is an area which industrial hygienists within the pharmaceutical industry can provide input and again showcase significant value to the organization by demonstrating definitive risk management principles.

NOTES

1 ISPE Good Practice Guide. "Containment for Potent Compounds." (1st Ed.). 2022.
2 ASHRAE 110–2016. Methods of Testing Performance of Laboratory Fume Hoods.
3 ISPE Good Practice Guide. "Assessing the Particulate Containment Performance of Pharmaceutical Equipment." (2nd Ed.). 2012.
4 EN 689. (2018). *Workplace Exposure - Measurement of Exposure by Inhalation to Chemical Agents - Strategy for Testing Compliance with Occupational Exposure Limit Values.* Bruxelles, Belgium: European Committee for Standardization EN 689:2018.

8 Risk Treatment
Administrative Controls and PPE

8.1 INTRODUCTION

Engineering controls are the primary means by which a risk management team effectively treats exposure risks to employees (Figure 8.1). Yet even the most well-designed hood, containment device, or piece of equipment is only effective if used properly and as intended. The way operators utilize devices is a component of administrative controls. Within the hierarchy of controls, administrative controls are below engineering controls in terms of effectiveness. Despite this widely held view, administrative controls are a critical component to the effectiveness of controlling employee exposures. At the bottom of the hierarchy of controls lies personal protective equipment, or PPE. PPE is ubiquitous throughout every industry, including the pharmaceutical sector. Widely considered to be a "last line of defense" by health and safety professionals, PPE is often viewed by operators as the primary (and sometimes only) means of keeping them safe in the working environment. This disconnect in understanding the hierarchy of controls is not unique to the pharmaceutical sector and has existed for as long as modern industry has existed. The detailed selection of required PPE and which administrative controls to enforce for a process are often guided the control banding scheme, which in turn is often dictated by OEB of the material being handled. Modern control banding schemes will list not just required engineering controls, but administrative controls and PPE as well.

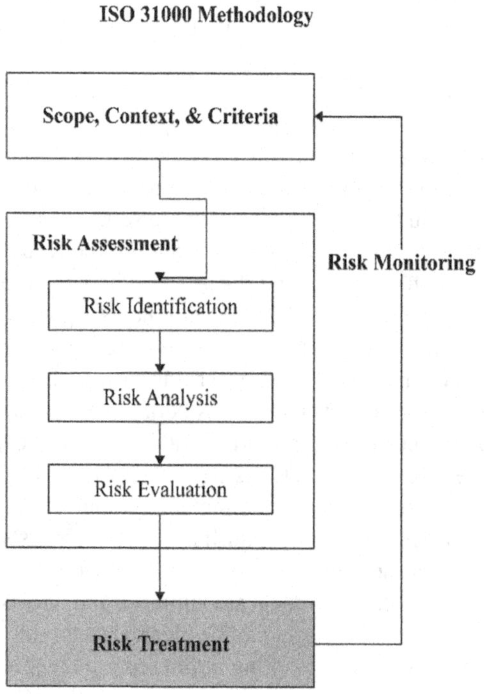

FIGURE 8.1   One of the final steps of the ISO 31000 risk management flow is risk treatment.

DOI: 10.1201/9781003273455-8

In this brief chapter, we will cover aspects of both administrative controls and PPE and how they can be used together with engineering controls to provide the safest working environments for workers by reducing their risk to exposures.

8.2 ADMINISTRATIVE CONTROLS

8.2.1 STANDARD OPERATING PROCEDURES

In the modern pharmaceutical industry, standard operating procedures (SOPs) are ever present in every facet of the industry. SOPs exist for processes within manufacturing, quality, safety, engineering, maintenance, and every other conceivable department. The intent of SOPs is to present the correct steps of performing a task so that they can be repeated without worry of a deviation. In the industrial hygiene realm, these documents are particularly important when they involve the proper use and maintenance of engineering controls. A properly authored SOP will showcase how to work with IH equipment so as to make proper connections, maintain airflow, or to simply operate equipment safely.

To be effective, SOPs must be sufficiently detailed to show how to perform the task or use the equipment properly without additional guidance, but not so excessively detailed that reading the document is time prohibitive. This is a fine line to walk for the document author. Historically, SOPs were abbreviated documents that did not supply significant information. In situations such as these, the document itself was ineffective and users learned how to conduct tasks from seasoned operators rather than the SOP. Or worse, users would use trial and error to figure out a way to make it work best for them, leading to multiple ways to perform a task (a significant quality and safety risk).

Modern pharmaceutical SOPs are no longer abbreviated or vague, but they are also not tomes through which a user must toil to learn anything.[1] The use of clear, concise language in the format of bullet points is frequently used. Moreover, SOPs are no longer just text. The inclusion of pictures, flow diagrams, piping and instrumentation diagrams (P&IDs), and other images significantly improve user context while minimizing text. A common use of images in safety-related SOPs is found in lock out/tag out documents, which showcase exactly which switches and electrical sources to utilize for safe handling.

An often murky point to consider is deciding who is the owner of SOPs within an organization. In the past, each department wrote their own document, perhaps based off a common template, and was responsible for its upkeep and ensuring its use. Today, most SOPs are indeed authored by a particular knowledgeable subject matter expert but they are reviewed and approved by members from various departments. These include members of quality, a separate technical reviewer, and others. The SOPs are typically housed by the quality department, but the responsibility for the documents review and enforcement lies with the authoring department. This method of reviewing documents by a collective and diverse team aligns with FDA requirements regarding quality.

For the practicing industrial hygienist, part of the risk assessment process involves reviewing SOPs associated with a particular process and observing how well the operators follow the provided steps. It is not uncommon to discover shortcuts being taken because the overarching SOP is unclear or lacking detail. Another point of providing value is for the hygienist to assist in reviewing or authoring SOPs for new equipment. This is particularly prudent given that he or she will be considered the subject matter expert for any containment equipment. Yet it is especially important to remember that individuals who are overly familiar with a process or equipment are likely to provide simplistic guidance because they assume the users have the same level of knowledge as them. This is another reason why the documents are authored and reviewed by a team rather than a single individual. Additionally, further insight can be gleaned by giving drafts of updated SOPs to line workers and asking them to follow it. Seeking unfiltered guidance from line workers may take additional time to conduct, but the results are often well worth the effort. The authoring of clear, effective, and comprehensive SOPs is the first step toward ensuring employees operate equipment and perform processes correctly. These documents must be reviewed periodically, and any changes should be thoroughly documented.

8.2.2 Training

Once an SOP is written and approved, the next step is to provide training on said document. Training is perhaps the frequently instituted administrative control as it has roots in regulatory requirements from OSHA and the FDA. Seasoned veterans of the pharmaceutical sector can vouch that there is no shortage of required trainings, which must be performed annually. The goal of training is to ensure that end users receive the same information and clarify any potential misunderstandings.

The manner in which trainings are given can vary between three general delivery mechanisms: computer based, classroom lecture, and hands on. Unquestionably, the most common method of training delivery is via computer-based methods. Computer-based methods utilize a wide variety of modern technology but typically involve the use of videos or point-and-click interactions. For SOPs, especially those pertaining to safety topics or processes, computer-based training involves reading the document and acknowledging that the user understands the information. Importantly, the acknowledgment does not necessarily require proof of competency or understanding. The more rigorous computer-based training for SOPs may involve a short multiple-choice quiz. The second method of training delivery is a classroom lecture. Such training sessions are often performed for instances where recipients may have questions and need to interact with an authoritative personality (i.e., the trainer). The content can encompass multiple topics and is often truncated and presented in the format of a PowerPoint presentation. The final form of training delivery is the hands-on approach. This involves showing the recipient exactly how to perform a given task in a controlled environment. Employees undergoing this form of training get to physically interact with equipment or perform a particular task. Hands-on training is unquestionably more time intensive and expensive to perform.

Of the three training delivery mechanisms, computer-based trainings are far and away the most common, especially for safety-related SOPs. An organization can easily have dozens if not hundreds of these documents which require an employee to review. Most of these are simply "read and understand" trainings, but the document author can decide if a competency exam is required (unless there is a regulatory requirement to showcase competency). The reality is that there are simply too many SOPs and too many employees to give the trainings in any other capacity. And this training method often holds true for new processes, changes to old processes, and the installation of new equipment. But there is debate in the field as to the effectiveness of computer-based safety training.

In 2006, a meta-analysis by Burke and colleagues was published regarding the methods of previously documented safety training and their effectiveness.[2] In their report, they reviewed more than 90 published studies on health and safety training and were able to categorize the types of training into the three groups we have already mentioned. Their analysis showed that hands-on training in which the trainee was highly involved with the training session produced greater knowledge retention and also reduced accidents, illnesses, and injuries than the other passive training methods (i.e., computer-based). Many health and safety professionals would agree with these conclusions based on subjective observations.

Unfortunately, the hands-on and engaging methods of training are very time consuming, and not just to execute but also to prepare. Any and all training material must be crafted and approved by the primary department but also the quality department. Furthermore, the time needed to perform such trainings is another critical time component. If multiple hours are needed to fully realize the training, that is a lot of time that is being lost to production. As always, there is a business element that must be considered and weighed prior to implementation

Yet many organizations are beginning to realize the value of in-depth, hands-on training as the primary means of conveying knowledge. This is especially true for pharmaceutical companies that handle HPAPIs. Often the training requirements include elements of all three categories: read and understand the SOP, receive a lecture component, and perform a skills-based assessment. The need for this level of training with potent compounds is obvious: even the smallest exposure to such materials can be detrimental to the health and well being of the employee. But these same organizations are beginning to implement similar levels of training for other direct material handling endeavors or for operations in which there is appreciable risk of exposure due to mechanical failure. For example, the process of connecting an FIBC is one which as the potential for exposures that exceed the OEL in the case of a connection failure simply due to the volume of materials handled. Even though the materials are often OEB 1 substances, the risk is prevalent enough to warrant a hands-on training approach. This is yet another aspect of how the industrial hygienist can showcase value to the organization. During a hands-on training session, the hygienist can perform sampling to determine the overall effectiveness of the operator. If the sampling data align with previously obtained data points, then the operator can be cleared for the operation using objective data rather than subjective clearance from a supervisor or trainer.

Regardless of how it is executed, training on new equipment and processes is a critical administrative control for the industrial hygienist. Proper training is vital to ensure that the variability in exposures is minimized as much as possible. For new equipment or modified processes, all training materials must be updated, and the revised content must be distributed to all affected members of the SEG. From a time and cost perspective, it is understandable why many organizations continue to favor passive training methods rather than hands-on training, but the latter format will provide a better long-term investment with fewer incidents and quality deviations

8.2.3 Worker Rotations

An oldie but goodie, many organizations use worker rotations through tasks and processes to minimize their overall exposure levels. In essence, by limiting the number of hours a worker is exposed to a substance provides more time for the employee's body to clear it from his or her body and also to sustain a lower chronic dose. This is a widely adopted practice that continues to this day in every industry.

One of the few drawbacks to this administrative control is that by rotating employees through various processes in a plant, it increases the total number of people who can be potentially exposed. In other words, worker rotations can increase the size of the SEG for a process. While expansion of an SEG is not automatically a negative outcome, the hygienist and the production department must be aware of the downstream effects should exposures for the SEG be deemed of an unacceptable risk. This would include a larger number of personnel to be included in the respiratory protection program, more PPE to be purchased, and a larger health surveillance effort. Even with these potential drawbacks, the pros typically outweigh the cons from a risk perspective.

8.2.4 Preventative Maintenance (PM)

For industrial hygiene risk treatment projects, the inclusion of preventative maintenance is sometimes an afterthought. A significant amount of work often goes into the design and implementation of a new control device, and good bit of work is done to ensure SOPs and training materials are updated as well. But all too often, the creation of preventative maintenance tickets with the engineering or maintenance departments slips through the cracks. Effective preventative maintenance ensures that equipment is running in the same (or at least similar) manner as when it was installed. For equipment such as containment devices, this means the containment performance continues to meet the needed requirements.

Preventative maintenance comes in many forms. It can include measuring SP within an LEV system, routine cleaning of duct work that serves dust collectors, changing filters, performing oil changes on motors, evaluating belts and bearings on exhaust fans, measuring total air flow in a system, and monitoring air pressures in a compressed air line. These examples are just a drop in the bucket of the types of actions that can be completed as a PM. The creation of a PM within an organizational maintenance system usually creates a trackable action, or a ticket, which is then assigned to a team member for completion by a specific date. The members of the maintenance team are often an excellent resource to utilize when determining what types of PMs should be created and how often they should be performed.

And again, the industrial hygiene group can provide a means to streamline the implementation of PMs and their subsequent completion. For many pieces of equipment that play a role in risk treatment for the industrial hygienist, several types of measurements will already be taken which may relate directly or indirectly to them. Such measurements can be routinely obtained by the IH group in an effort to support the maintenance department. For example, the practicing hygienist is often familiar with performing duct traverses or measuring SPh for a given hood. Such measurements are typically taken upon installation (the SAT) and can serve as the baseline for performance of the equipment. The IH group can offer to acquire the same measurements on pre-determined schedule as part of the PM for the equipment. In this manner, the IH gets to see first-hand how the equipment is performing while also serving to complete a routine PM and provide assistance to a partnering group. Moreover, the acquired data can be compared to the baseline data, which should be accompanied by sampling data. Over time, enough data points can be acquired to generate a control chart. The utility of the control chart is such that it provides time-stamped data to show equipment performance over time, which can be used to adjust the PM schedule as necessary.

The importance of a good PM program cannot be overstated. Far too frequently equipment and processes are installed or implemented without any follow up or maintenance to ensure their longevity. Undoubtedly, this is a primary reason why equipment breaks without warning or utility costs seem to run significantly higher than they should. PMs take time and money to implement, but in the grand scheme of manufacturing and risk management, it is far more prudent to know that you will have to replace a $150 part on a piece of equipment, even if it may not be broken, instead of having to unexpectedly pay $20,000 to have it replaced when said part breaks down. PMs ensure the equipment and the production schedule do not significantly deviate.

8.2.5 SIGNS AND LABELS

A final administrative control we will touch on is signs and labels. As with SOPs and training, signs and labels are prominent fixtures in any manufacturing area. Employees are often showered with a wide array of signs in the workplace. In the IH and risk treatment realm, such postings are an important hazard communication tool. Employees who work with a particular substance for long enough can easily become complacent to the hazards associated with it. Perhaps even worse, workers may not be fully aware of the hazards in the first place and therefore may not realize the importance of all the other controls surrounding the process.

Signs and labels can convey a wide range of critical information that the IH needs the operators to know. This can include relaying the required PPE for the task, the proper ergonomics to use for a task, potential physical hazards of a substance (such as flammability or combustible dust), or potential health hazards of a substance (such as carcinogenicity, teratogenicity, or target organ effects). Signs can even indicate instructions for separate actions, such as what to do in case of emergencies (examples of this can include safety shower and eye wash stations or explaining what audible/visual alarms may represent). Since hazard communication is a critical component of industrial hygiene, the appropriate use of signs and labels must also be utilized effectively.

8.2.6 ADMINISTRATIVE CONTROLS – SUMMARY

Administrative controls are a critical piece of the hierarchy of controls and to the process of risk treatment. Without proper administrative controls, even the greatest containment device or LEV system will fail every time it is used because the systems will not be used correctly. Unfortunately, administrative controls are often glazed over during the risk treatment stage, taking a back seat to the engineering control design. Well-crafted SOPs with input from all stakeholders will go a long way to assisting operators on the floor, but these documents only work if appropriate training on the SOPs is performed. Passive training methods (such as computer-based training) remain the most popular training technique today out of necessity. Most pharmaceutical organizations have too many SOPs to be able to perform adequate in-person training. But research has shown that interactive hands-on training results in fewer safety issues and deviations, making it a more effective pedagogical tool. Many companies have begun utilizing this method of training for the higher-risk processes or those in which higher-banded materials are handled (such as ADC-handling facilities). Worker rotations continue to be a useful means of reducing employee exposure while keeping them engaged with different activities to prevent monotony. Preventative maintenance is an important aspect to keeping equipment running after installation, and the IH group presents a unique ability to contribute to these and provide additional value. Finally, signs and labels are an integral component of hazard communication and should be used judiciously in prudent areas. Administrative controls have the potential to significantly reduce employee exposures and are an often-underrated resource.

8.3 PERSONAL PROTECTIVE EQUIPMENT (PPE)

Personal protective equipment (PPE) is the lowest rung in the hierarchy of controls and considered the least effective. This consideration is based on the fact that all other forms of control (elimination, substitution, engineering, and administrative) aim to prevent exposures from occurring to the operator. PPE, on the other hand, is a last-resort defense for the employee to prevent unintended spills, splashes, or airborne emissions from affecting the employee. PPE is indispensable within the pharmaceutical industry – or any industry, for that matter. PPE is so universally accepted and pervasive that whole companies exist which are dedicated to supplying a single type of PPE. There exists PPE for protecting all body parts to which injuries can occur – eyes, ears, hands, back, chest, feet, even the entire body. Moreover, specific PPE exists for protecting against specific hazards, such as gloves for protection against arc flash dangers. Within the pharmaceutical industry, and from an industrial hygiene perspective in particular, we will focus specifically upon PPE for protecting the lungs, eyes, hands, and body. While hearing protection also falls under the purview of industrial hygiene, this book has focused exclusively on exposures to airborne contaminants. Continuing in this vein, we will exclude hearing protection from the discussion, although in practice it must always be considered.

8.3.1 Respiratory Protection

The most prominent PPE program within any organization is (usually) the respiratory protection program. The reason for the prominence is that there are strict regulatory requirements for a respirator protection program, and consequently, they receive additional scrutiny during OSHA and corporate audits. Additionally, the most likely route of exposure for employees during a work shift is via inhalation. It therefore makes intuitive sense to have a well-developed respiratory protection program. The specific requirements for a compliant and effective respiratory protection program in the United States are outlined in 29 CFR 1910.134.

The crux of respiratory protection is found in the equipment themselves. Respirators come in an assortment of types and can be classified in several ways: tight-fitting or loose-fitting; half-face or full-face; particulate filtering or vapor filtering. The means to mix and match a respirator are numerous and very much dependent on the specific situation that it calls for. However, of paramount importance is the assigned protection factor (APF) of a respirator. The APF is an assigned value that represents the workplace level of respiratory protection that a respirator can be expected to provide when used properly. In the United States, these values are assigned by OSHA or by the PPE manufacturer (with test data to back up the APF claim). The APF values associated with some of the more common respirator types are shown in Table 8.1. All of the respirators listed in the table are commonly utilized in the pharmaceutical industry depending on the process involved. For instance, N-95 respirators are utilized in areas that are dusty and where the airborne levels are not extremely high, yet they are not used for vapor protection. This is because N-95 respirators are not able to filter out gases or vapors, only particulates. In contrast, the half-face respirator can be used for protection

TABLE 8.1

Commonly Assigned APF Values to Various Respirator Types

	Assigned Protection Factor (APF)	Fit Test Required
N-95 (dust mask)	10	Yes
Half-face respirator	10	Yes
Full-face respirator	50	Yes
Supplied air hood	1,000	No
Powered air purifying respirator (PAPR)	1,000	No

against either form of airborne contaminant (depending on the cartridge used). But half-face respirators have the same APF as N-95 respirators and do not offer eye protection. Full-face respirators do offer eye protection and have a higher APF as well, yet these RPE are larger, have more parts to clean and maintain, and are significantly less comfortable than their half-face counterparts. All three of these respirators must also be fit tested on an annual basis, and a physician must give clearance for the employee to use the respirator upon hiring. Supplied air hoods rely on a pressurized source of breathing air to be connected to a loose-fitting hood or shroud that covers the employee's head. A continuous stream of clean, unadulterated air is provided within the shroud. Since this is a positively pressured environment, it has a much higher APF than any of the tight-fitting alternatives. Additionally, no fit testing is required. However, the operator is restricted by the length of hose that supplies the air. It also potentially serves as a trip hazard when operations are underway. A similar alternative is the PAPR, which is a battery-powered respirator unit that supplies filtered air through a similar shroud used for supplied air respirators. In essence, the PAPR offers the same advantages (APF, absence of fit testing, etc.) as supplied air but without the burdensome tether. PAPR units are commonplace in areas where HPAPIs are handled on any scale as they offer the highest APF with the greatest mobility and flexibility.

The requirement to know the APF of a respirator is paramount for the industrial hygienist as it is used to calculate the maximum use concentration (MUC) of a substance being handled. The MUC is the highest concentration of a substance to which an employee can be exposed and still expect to be fully protected by the respirator. The MUC is found by multiplying the OEL of a substance by the APF of the respirator. For example, if a substance with an OEL of 400 µg/m³ was being handled and an N-95 respirator was chosen as the RPE for the task, the MUC for the process would be 4,000 µg/m³:

$$MUC = APF \times OEL = 10 \times \frac{400 \ \mu g}{m^3} = \frac{4,000 \ \mu g}{m^3}$$

But how is this used in practice? Most pharmaceutical organizations are conservative and aim to adequately protect employees during the highest exposure periods, which would be during the task-handling activities themselves. In this respect, the STEL limits are frequently used to derive the most prudent respirator type for the process. Moreover, when analyzing the sample data for an SEG, the industrial hygienist is able to determine if the chosen respirator is in fact the most appropriate for the task and the level of exposure. This can be done using BDA and assessing whether the 95th percentile of the task data exceeds the MUC for the chosen respirator. If it does, then the respirator is inappropriate for the task being performed and a more rigorous one should be chosen. Additionally, this would mean that the airborne levels are in fact very high and efforts should be undertaken to rectify this issue. In our example above, if the 95th percentile exceeded the MUC, in practical terms that means the airborne levels would exceed 4,000 µg/m³, which is more than ten times the allowable limit. This situation would be an immediate red flag and begin a more in-depth root cause analysis to explain the high levels.

Significant challenges exist with the implementation of respirators in the pharmaceutical realm. While the first challenge is deciding on a suitable respirator for a given task, a second and significant challenge is to ensure that employees wear the respirators appropriately. It is a common occurrence for employees to neglect wearing RPE when it is required or to grow facial hair at some point during the year (facial hair breaks the seal around a tight-fitting respirator, negating its effectiveness). These instances occur even if the requirement of respirator use is mandated by administrative controls (SOPs, training, etc.). Complaints of discomfort are common. There are many reasons why respirators are not diligently worn by employees, but it is up to their respective line managers to ensure the required RPE is being worn. A respirator is only effective if it is worn and worn properly. Another challenge is ensuring that fit testing occurs on a regular basis and that operators do not lapse in this requirement.

8.3.2 HAND PROTECTION

Hand protection is accomplished by using gloves. Gloves are so common in every industry that to not have them available would be unfathomable. They are available from numerous manufacturers in assorted sizes and for a wide variety of manufacturing situations, including protection from physical, chemical, biological, electrical, thermal, and mechanical hazards. For the purpose of brevity, we will focus on glove selection as it pertains to employee protection against chemical exposures.

In the United States, the requirement of an employer to offer hand protection is a regulatory requirement (29 CFR 1910.138). Yet the requirements are brief, stating that employers must select and require the use of appropriate hand protection and that the selection of said PPE will be based on an evaluation of the glove performance characteristics for a given hazard. It is incumbent on the employer to properly select a glove for a given process. When viewed through an industrial hygiene and risk treatment lens, this means selecting a glove which offers the greatest protection against chemical exposure. Historically, this meant evaluating how well a glove withstood exposure to liquids (such as solvents) by measuring permeation and degradation rates. Permeation is a process by which a substance diffuses through the membrane of a material without the need of a tear, rip, puncture, or pinhole. There are several standards which outline how to measure permeation rates, such as ASTM F379 and EN-16523. Permeation rates are typically measured by placing a membrane (glove material) in a sealed test chamber and placing a challenge chemical, such as a solvent, on one side and measuring how much time transpires before the liquid is detected on the other side. Simultaneously, the glove is measured for how well it stands up or "degrades" upon exposure to the chemical. Such tests are performed under ideal laboratory conditions.

Glove manufacturers typically compile their test data into easy-to-read charts that are broken down by glove material and compared against a list of challenge solvents. The most common glove materials are typically listed, including neoprene, nitrile, viton, butyl rubber, and polyvinyl alcohol (latex gloves used to be a common glove choice, but an increasing percentage of the worker population exhibits a contact allergic reaction to latex, and consequently, it has been phased out of most manufacturing areas[3]). It is these glove charts that are often the information source for glove selection. Yet there is concern such rationale only paints part of the picture for a realistic selection. It has been suggested that additional confounding factors such as the use of mixtures, temperature, and also physical stress (bending, stretching, and general pressure) applied to gloves should also be factored into glove testing and selection.[4] The addition or inclusion of such factors into the testing standards has also been suggested.[5]

Inclusion of such confounding factors into appropriate decision making is not disputed by EHS professionals, but how to include them into the process remains a challenge. Without any objective standards utilizing their inclusion or any test data, the organization is left to make an informed choice using subjective assumptions about a process and how it will impact glove durability. Furthermore, the test standards apply to liquids and gases, with little attention given to solids in solution. In the pharmaceutical industry, such concerns are highly prevalent, especially when APIs are involved. While there have been independent laboratory assessments of permeation rates of solutions containing hazardous agents,[6] more work is still needed to ensure that gloves are offering the greatest amount of protection to employees during specific tasks.

Like with respiratory protection, there are unique challenges to ensuring hand protection is done correctly. While employees will almost instinctively use gloves in any setting where chemicals are used, it is imperative that they use the correct gloves. As mentioned above, different gloved materials perform very differently with an array of chemicals. Usually, the desired glove for a process is stated in an associated SOP, but this is only one part of enforcement. All too often supply managers will seek out gloves which have a lower unit cost to save money, thinking that any glove made of a particular material is the same from every manufacturer. Unfortunately, this is not the case, as there are different formulations and thicknesses to consider. This leads to a false sense of security and could potentially put employees at risk of exposure. Some SOPs may also require employees to utilize two layers of gloves, often referred to as "double gloving". The reason for this requirement can stem from a glove's performance with a certain chemical or from a quality perspective to avoid cross-contamination. Yet a more common industrial hygiene reason for double gloving is to provide a layer of protection to the operator while the soiled outer glove is removed. Such protocols are common with potent compound handling.

8.3.3 BODY PROTECTION

When considering routes of exposure to employees, general body exposures do not usually contribute to a significant extent. However, total body protection is still paramount in the pharmaceutical industry. The primary drivers for providing body protection are to provide employee safety from acute hazards (splashes and spills of liquids and powders), quality concerns, and the possibility of material being "tracked" through a facility or contaminating an employee's personal garments.

There are two common methods of body protection: reusable jump suits and disposable chemical protective clothing (CPC). Jump suits are frequently used whenever the materials being handled are primarily powders and are less hazardous in nature (OEB 1 materials, for example). Liquids are generally not suitable for jump suits as they are absorbent. Splashes and incidental contact with liquids can be absorbed where they will quickly pass through the material and remain in contact against skin. While this is the general consensus, there are instances wherein jump suits are used with liquid handling processes. In such instances, well-established emergency protocols are in place to circumvent any potential mishaps.

A significant drawback to the use of jump suits is the requirement to have them laundered or cleaned. Such tasks are generally performed by third-party groups, and they must be fully informed of what the potential contaminants are on the uniforms. Failure to do so is not only unethical but illegal, since it can potentially put the employees of the laundering service at risk of exposure. Depending on the size of the facility, the number of uniforms to get laundered can be quite large. This leads to additional recurring costs which must be accounted for. Another drawback to the use of jump suits is their tendency to have powdered materials "stick" to them. Powders and dusts often adhere to fabric materials, making them a potential secondary source of exposure. Simple wiping efforts of the garments usually do very little to remove significant quantities of material, thus necessitating the laundering requirement.

In contrast, disposable CPC provide convenient solutions to the issues posed by fabric jump suits. Disposable options are often comprised of densely packed materials, such as polyethylene which makes them highly resistant to liquids and powders. They are also frequently tear resistant, making them more durable. The long list of substances that disposable CPC can withstand makes them ideal in the pharmaceutical industry. Indeed, they are frequently used in a variety of manufacturing and quality arenas. After a particular task or process is completed, the operator can simply properly doff the garment and dispose of it, eliminating the need for laundering services and reducing risk in the process.

But just like with gloves, not all manufacturers of disposable CPC make equivalent products. Such garments are subject to the same permeation risks that gloves are. Depending on the nature of the chemical being handled, over time the substance can diffuse through the clothing and present an exposure risk. It is important to understand these permeation rates, especially since it is far more difficult to change garments than to change gloves when they become soiled. In addition to permeation rates, there also exists the need to know about penetration rates. Penetration is the tendency of chemicals to find their way through incidental gaps in the material. This most frequently occurs along the seams of the clothing where they are sewn together. When selecting a disposable CPC, it is often important to note how tight the seams are. Many are sewn together very tightly, and still others are manufactured using a heat-sealing process, thus all but eliminating the seams altogether. Because manufacturing methods and materials vary, it is important to select a desired type of disposable CPC and ensure it is always ordered.

There are other drawbacks to the use of disposable CPC. The first issue is the need to continually order and stock the garments. During the COVID-19 pandemic, many organizations encountered the same problem with the supply chain. On occasion, vendors were out of stock of the desired garments with long lead times. Yet even if there are no issues with the supply chain, the recurring cost of purchasing the garments must be taken into consideration and applied to the PPE budget. And since the garments are disposable, they increase the volume of waste. Disposable CPC are often treated as hazardous waste since they have the potential for a wide number of chemical contaminants, which may exhibit one or more of the EPA characteristics of hazardous waste. But even if they are not disposed of as hazardous waste, they will still go through whatever the typical means of waste disposal are for a facility. Such garments only add to the waste produced by a site and in today's sustainability-driven culture there is often reluctance to adopt another route which adds to the waste streams.

Another and perhaps more important issue stems from the garment's ability to protect against chemical exposures. Disposable CPC are often so tightly manufactured that they also prevent heat from escaping. When fully suited, an employee is likely to get very hot very quickly. It is a common site to witness operators emerging from such garments completely soaked with sweat. Since manufacturing environments can also be very warm, this only adds to the interior temperature of the garment. The danger of dehydration or even heat stress is a real risk for many operators. It is important to ensure that such working areas are sufficiently cooled to prevent such instances from occurring. Industrial hygienists with experience in thermal stressors can provide additional insight to this aspect.

In spite of these drawbacks, the use of disposable CPC is the most common form of body protection offered to employees, especially when handling materials in higher level exposure bands. Their convenience, relative affordability, and capacity to reduce multiple avenues of risk (chemical exposure and quality) make them a commonly observed sight in all areas of the industry.

8.3.4 Eye Protection

PPE for eyes is another significant requirement. In the United States, eye protection has its own regulatory standard, 29 CFR 1910.133. The primary requirement stemming from this standard is that all eye protection must meet the ANSI Z87 consensus standard and that the appropriate eye and face protection must be offered to match the given workplace hazard. In the pharmaceutical industry, the most common eye protection is safety glasses which are impact resistant (for protection against flying objects) but which also contain side shields for additional protection against chemical splashes. Some users go a step above and utilize impact resistant chemical goggles which form a tight seal around the eyes for significantly better chemical protection.

The requirement to wear eye protection at all times is often enforced at pharmaceutical sites. The requirement not only applies to employees actively engaged with material handling but bystanders or passersby. There are numerous administrative controls that apply to eye protection, including

SOPs, training, and numerous signs and labels throughout a facility. Since eye protection is relatively inexpensive, some operators dispose of their eye protection at the completion of a shift. This practice is rather wasteful and costly, especially since eye protection can be easily decontaminated and reused. For operators who wear corrective lenses, organizations typically offer PPE with prescription corrective lenses in them. Standard corrective lenses are not classified as PPE and thus cannot be worn in a manufacturing environment.

8.4 SUMMARY

Administrative controls and PPE are often-underrated tools for reducing exposures to employees. The goal of administrative controls is to provide instruction and education to the operator so he or she can perform their tasks as intended. When executed appropriately, administrative controls ensure the processes are performed the same way every time and without incident, thereby ensuring quality for the product but also the safety of the personnel, which is the most important aspect. Administrative controls serve to prevent exposures from occurring. In contrast, PPE serves as a last line of defense if unintended exposures occur. The specific administrative controls and PPE for a task are typically provided by the organizational control banding scheme, but the implementation of such controls is the responsibility of the site. The practicing industrial hygienist plays a key role in reviewing administrative controls (SOPs, training, etc.) and helping to select PPE that matches the task and nature of the hazards present.

NOTES

1 O'Leary, T. "Creating effective standard operating procedures". ISPE Blog entry, January/February 2021. https://ispe.org/pharmaceutical-engineering/january-february-2021/creating-effective-standard-operating-procedures.

2 Burke, M., et al. (2006). Relative Effectiveness of Worker Safety and Health Training Methods. *American Journal of Public Health, 96*(2), 315–324.

3 Reiter, J. E. (2002). Latex Sensitivity: An Industrial Hygiene Perspective. *Journal of Allergy and Clinical Immunology, 110*(2), S121–S128.

4 Klingner, T. D. and Boeniger, M. F. (2002). A critique of assumptions about selecting chemical-resistant gloves: A case for workplace evaluation of glove efficacy. *Applied Occupational and Environmental Hygiene, 17*(5), 360–367.

5 Banaee, S. and Que Hee, S. S. (2020). Glove permeation of chemicals: The state of the art of current practice - part 2. Research emphases on high boiling point compounds and simulating the donned glove environment. *Journal of Occupational and Environmental Hygiene, 17*(4), 135–164.

6 Oriyami, T., et al. (2017). Evaluation of the permeation of antineoplastic agents through medical gloves of varying materials and thickness and with varying surface treatments. *Journal of Pharmaceutical Health Care and Sciences, 3*, 1–8.

Index

Printed in the United States
by Baker & Taylor Publisher Services